Swc

Knives, Swords, and Bayonets:
A World History of Edged Weapon Warfare

by Martina Sprague

Copyright 2013 Martina Sprague

All rights reserved. No part of this book may be reproduced in any form or by any means, electronic or otherwise, without the prior written consent of the author.

Acknowledgements:

Front cover image pictures tombstones of Roman cavalry with spear and sword. Image source: C. Iulii Caesaris belli gallic, reproduced under Wikimedia Commons license.

Back cover image pictures coin portraying Greek mythological hero Ajax, son of Oileus. Image source: Marie-Lan Nguyen, coin bequeathed by Mrs. Adra M. Newell to the Metropolitan Museum of Art, reproduced under Wikimedia Commons license.

Image source for horse logo (slightly adapted) on back cover: CoralieM Photographie, reproduced under Wikimedia Commons license.

Swords and Warfare in the Classical World

TABLE OF CONTENTS

Some Notes about the Knives, Swords, and Bayonets Series	4
Introduction	10
Bronze, Iron, and Steel	19
Sword Training	32
Military Organization and Battlefield Tactics	36
War Wounds and Protective Armor	55
Concluding Remarks	63
Notes	68
Bibliography	85

SOME NOTES ABOUT THE KNIVES, SWORDS, AND BAYONETS SERIES

Knives, Swords, and Bayonets: A World History of Edged Weapon Warfare is a series of books that examines the history of edged weapons in Europe, Asia, Africa, the Americas, and the Middle East and surrounding areas before gunpowder increased the distance between combatants. Edged weapons were developed to function in foot or mounted combat. The primary battlefield function often determined the specific design of the weapon. In poorer societies the general populace frequently modified agricultural tools into weapons of war. The techniques for employing these tools in civilian life translated into viable methods of combat. When the advent of firearms made certain edged weapons obsolete, close range combat continued to rely on foot soldiers carrying knives and bayonets as sidearms to modern artillery weapons. But even in ancient times edged weapons were seldom the primary arms, but were frequently employed as sidearms to long range projectiles. Rebel fighters of Third World countries have likewise used edged weapons extensively in near modern and modern wars.

The Knives, Swords, and Bayonets series of books takes a critical look at the relationship between the soldier, his weapon, and the social and political mores of the times. Each book examines the historical background and metallurgic science of the knife, sword, or bayonet respectively, and explores the handling characteristics and combat applications of each weapon. The author suggests that the reader make specific note of how battlefield need and geography influenced the design of the weapon, the type of warfare employed (guerrilla, rebellion, chivalry, pitched battle, skirmishes, mass armies, etc.), and the type of armor available to counter the blow of a knife or sword.

The historical treatment of edged weaponry could fill volumes. Because of the vastness of the subject, certain restrictive measures had to be applied in order to keep the series within a reasonable length while still giving adequate coverage. For example, the author has chosen to cover Chinese and Japanese but not Korean sword history. Every reader is thus bound to find some favorite details omitted. While many treatments of the subject focus exclusively on the technological aspects of weapons, this series also considers the political climate and the environmental or geographical factors under which the weapons evolved. Moreover, every culture, western or non-western, employs a number of subtleties that are exceedingly difficult to understand fully, unless one has spent time living in and studying the specific culture. The same can be said for every subculture (a culture within a culture), such as a military organization. The reader is reminded that, unlike science which is mathematically precise, history offers a broad range of perspectives on every issue.

The narrative the author has chosen to write portrays the development and dynamics of edged weaponry from ancient to modern times, including the soldier's training and his view of military service. The close relationship between military and political or social history also spurred the author's desire to examine the carry of edged weapons as symbols of military rank and social status. Rather than covering battles in their entirety, the author has elected to illustrate bits and pieces of particular battles that exemplify how the weapon in question was used. The book series comprises ten books arranged by weapon type, geographical area, or time period, and is designed to introduce the reader to the great assortment of edged weaponry that has been used with varied success in most regions of the world. Each book in the series is an entity in itself. In other words, it is not necessary to read the books in any particular order. Hopefully, the series will provide the reader with a solid foundation for continued study.

For her research, and in order to render an analysis that closely describes the dynamics of battle and the cultural aspects surrounding edged weaponry, the author has relied on a large number of primary and secondary source materials including historical treatises, artifacts located at museums, ancient artists' renditions of war in sculpture, paint, and poem, eyewitness accounts to the events in question, books, articles, documentaries, Internet resources, university lectures, personal correspondence, and direct hands-on practice with weapons in mock battles. Note that source material is often contradictory in nature. For example, swordsmen of the same era and geographical region frequently differed in their views with respect to the conduct of battle or the "best" type of sword or battlefield technique. The reader is encouraged to keep an open mind and consider the different possibilities that the soldier faced, and why he would emphasize a particular type of weapon or combat technique over another. The endnotes provide additional information, clarification, and exceptions to commonly propagated historical beliefs.

The author reminds the reader that despite their lethal features, edged weapons are not randomly chosen bars of steel that can cut and kill. The difference between victory and defeat often lies in the soldier's knowledge, skill, and fortitude; in how well he handles his weapon, but also in how well the weapon adheres to the laws of physics with respect to balance and motion. Studying metallurgic science is the key to understanding the relationship between the weapon smith and the soldier. The knife- or swordsmith thus carried part of the responsibility for the soldier's success or failure. Additionally, edged weapons were an integral part of the soldier's kit and often represented abstract qualities such as bravery and honor. By understanding the history of knives, swords, and bayonets, one will gain insight into the culture—the external and internal forces—that shaped the men who relied on these weapons in personal struggles of life and death.

Gladius and Spatha

Gladius and Spatha

Swords and Warfare in the Classical World

INTRODUCTION

Ancient Greece and Rome, also referred to as the Classical world, laid the foundation for Western civilization. Because of their geographical proximity to one another and to the Mediterranean Sea, the countries shared many similarities with respect to culture and weapons. Alexander the Great (356-323 BCE), king of Macedon, the northern part of ancient Greece, became conqueror of the Persian Empire in his mid-twenties, and allowed Greek culture to spread to western Asia.

In the fourth to third centuries BCE, war in Greece, as in many parts of the world, was a reality. Called upon when the political situation deemed it necessary to go to war, the Greek military consisted of heavily armored infantrymen fighting in phalanx formation and serving as citizen-soldiers, or hoplites. (The Greek word for weapon is *hoplon*, and can also mean protective armament such as a shield.) Although not professional soldiers *per se*—they returned after a campaign to their civilian occupations—due to the organization of the rank and file, the men were exceptionally disciplined, relying on group tactics rather than individual hand-to-hand combat. (The Spartans, by contrast—also called Lakedaimonians from the territory of Laconia in southern Greece—were educated from early childhood in the art of war. They endured strict training and can be viewed as an elite fighting force.) Xenophon (c. 431-354 BCE), an Athenian soldier and pupil and admirer of Socrates, "compared the arrangement of the hoplite line to the construction of a house," where the best and strongest materials are used for the foundation while the "mud bricks and timber are fillers placed in the middle." Those wishing to improve their individual skills took private instruction by paying the instructor a fee, and reaped the benefits of such training when their spear broke and they had to resort to the sword; a weapon requiring greater skill than strength to wield successfully.[1]

Unlike a sword with parallel edges which proved inefficient for cuts, the double edged *xiphos*, a leaf-shaped sword with the optimal mass near the center of percussion (the part of the blade that produces the least amount of vibration when struck), gave the hoplites the ability to thrust and cut through bronze armor. The *falchion*, likely descended from the Egyptian *kopis* or *khopesh*, with its heavy single edged blade and majority of weight toward the tip was also used and produced good slashing and hacking capabilities.[2] Other sword shapes existed as well, including a rather broad two-edged straight sword without taper used by the Greek infantry armies in the early period. However, it was the spear and javelin rather than the sword that remained the primary weapons of the Greek armies. Additionally, the hoplite considered his shield, which was reinforced with a bronze rim or thin bronze covering, his most important piece of equipment. The shield protected him against close combat weapons, such as spear and sword.[3] More importantly, it strengthened the formation and was used "for the common good of the whole line."[4]

Rome, perhaps the greatest power of the Mediterranean, differed from other civilizations most notably in "the notion of citizenship and the responsibilities and rights inherent in being a *civis Romanus.*" Italians had learned about constitutional rule from the Greek colonies and created the concept of a nation (in contrast to Carthage, for example, a city-state in North Africa with a government in the hands of a select body of aristocrats, which, according to military historian Victor Davis Hanson, "had not evolved beyond the first phase of Hellenic-inspired consensual rule"). Being Roman meant having citizenship. The Romans identified themselves by culture rather than by race, geography, or free birth. The idea of citizenship provided "a legal framework" for military operations. The men had "ratified the conditions of their own battle service," and created the sense that they had a stake in the outcome of a war.[5] When exploring the key characteristics of the Roman army, one might also consider the tribal origins of Rome which influenced the Roman method of war for several centuries and resulted in a hierarchical structure of society.

    The Romans were a pragmatic people who did not fail to adopt from other cultures the weapons or technology they found useful. In the early period, around 700 BCE, they constructed their swords of bronze. These had much in common with Greek swords, particularly with respect to the double edged xiphos.[6] The classic Roman short sword remained in use for centuries; however, while the Romans emulated many of the Greek weapons and armor, they also copied Hispanic daggers as well as straight Hispanic swords during the conquest of Spain in the third century BCE.[7] Refined with a lengthened point suited for stabbing through mail armor, these swords became known as the *gladius hispanicus*, or the Spanish sword, and were believed to be superior to the shorter Greek version. (The word gladiator, meaning swordsman, is derived from the gladius, which has almost come to symbolize the Roman warrior.[8]) Scipio Africanus, the Roman general and statesman who took New Carthage in 209 BCE, captured numerous Spanish swordsmiths, and forced them to manufacture swords for the Roman legionaries who used these weapons against the Macedonians in 200 BCE.[9]

Short gladius from the Legion of the Imperial Roman Army. Image source: Notafly, reproduced under Wikimedia Commons license.

While the swords of the Spaniards proved deadly whether used for thrusting or cutting, the Romans were critical of the swords of the Gauls, the northern neighbors of the Celt-Iberians (of the Iberian peninsula), considering them inferior, "being only able to slash and requiring a long sweep to do so."[10] The strong combinations of offense and defense employed by the Romans with their thrusting swords and shields efficiently protected them from the overhead blow of the Gallic sword.[11] What might it have been like, then, for a soldier in Classical times to engage an enemy in combat? According to Roman historian Titus Livius (59 BCE-17 CE), when a Roman and a Gallic soldier "took their stand between the two armies" and engaged in battle . . .

> . . . the minds of so many individuals around them suspended between hope and fear, the Gaul, like a huge mass threatening to fall on that which was beneath it, stretching forward his shield with his left hand, discharged an ineffectual cut with his sword with a great noise on the armour of his foe as he advanced towards him. The Roman, raising the point of his sword, after he had pushed aside the lower part of the enemy's shield with his own, and closing on him so as to be exempt from the danger of a wound, insinuated himself with his entire body between the body and arms of the foe, with one and immediately with another thrust pierced his belly and groin, and stretched his enemy now prostrate over a vast extent of ground.[12]

The ancient Roman and Greek authors were naturally biased in favor of their own disciplined and "civilized" societies, and may therefore have described the Gauls and Carthaginians in less than flattering terms. The people of Iberia, the region comprising modern Spain and Portugal, likewise suffered heavy losses in the early fights against the Romans, not necessarily because they were cowards and relied on skirmishes rather than pitched battle, but because they fought against a force that was better organized.[13] The strength of the Romans lay in the soldier's knowledge of his position in the formation; he was less likely to desert and more likely to protect his comrades. His position also allowed him to orient himself quickly and obey orders without first having to resort to a lengthy mental process.

Carthage, perhaps the greatest naval power in the Mediterranean in the third century BCE, was steadily expanding its empire over North Africa. Simultaneously, Rome was emerging as a dominant force and proved exceptionally successful in the expansion of its empire through the Mediterranean and Western Europe, including parts of North Africa and the Middle East. To contend with its enemies, the Roman yeomen, like the Geeks, had "voluntarily imposed civic musters." Although Hannibal Barca, the Carthaginian military commander during the Second Punic War of 218-201 BCE, was able to inflict severe losses on the Romans, they were determined to rid Carthaginian invaders from their soil.[14] The Greek historian Polybius (c. 200-120 BCE), who had spent eighteen years in Rome as a political hostage, had unique insights into Roman affairs. He wrote about the rise of Rome, compared the Empire's different enemies, and recorded that

Hannibal had equipped the Africans with arms in the Roman fashion, which had been captured in battle.[15]

The idea that one could become a citizen as long as one was Roman in spirit and expressed a willingness to pay taxes and protect Roman law had far reaching consequences: While few trained mercenary replacements were "available to Hannibal in the exuberance of victory, a multitude of raw militiamen recruits" were available to Rome "in the dejection of defeat." The Carthaginians would find it almost impossible to field troops as patriotic as the legions. The challenge Hannibal faced was not limited to winning at Cannae, but also included the elimination of more than a quarter of a million Italian farmers who were fighting for the promise of Roman citizenship. The Carthaginians, by contrast, fought not for Carthaginian citizenship but for hatred of Rome, for an opportunity for money and plunder, and for "freedom to govern their own affairs."[16] On the battlefield, the Carthaginian army rewarded individual kills rather than stressing rank and formation, and "the protection of one's comrades."[17]

The Roman Republic's geographical location centered on the Italian peninsula is important to how it developed militarily. The Roman army was mobile, well equipped, disciplined, and organized, and was able to deploy forces throughout the regions of the Mediterranean Sea. The legion, a loose conglomeration of companies, "each composed of two smaller 'centuries' of between sixty and one hundred soldiers" led by a centurion "who mastered the Roman system of advance and assault in unison," moved in "fluid formations" while the soldiers threw javelins and ran to meet the enemy head-on with their swords. The Roman success and the terror the enemy must have felt upon meeting the Roman advance, according to Victor Davis Hanson, lies "in the studied coolness" and the unmatched discipline of the Romans, and in the "predictability of the javelin cast."[18]

This book begins with an introduction to bronze, iron, and steel used in weapons and armor in Classical Greece, Britain, Gaul (western Europe, mainly in the France, Belgium, and Holland area), Rome, and Spain (or Iberia, the Spanish peninsula). Next it examines the type of training the soldiers encountered to prepare them for battle. This section focuses mainly on the training of the Roman warriors. The book then analyzes the military organization and battlefield tactics of the Greeks, Romans, and Celts. It ends with a discussion of war wounds and the protection that various types of armor afforded the soldier. The concluding remarks focus on the pragmatic way in which one viewed war in the Classical world.

Greek sculpture of winged youth with a sword, from the Temple of Artemis at Ephesos, c. 325-300 BCE, excavated by J. T. Wood for the British Museum, 1871. Image source: Marie-Lan Nguyen, reproduced under Wikimedia Commons license.

BRONZE, IRON, AND STEEL

Changes in battlefield tactics dictated the need for new innovations in weaponry. Around 1000 BCE, iron became the dominant metal for the construction of weapons, making it possible to manufacture blades of battle-worthy quality.[19] In the sixth century BCE, Hallstatt, a central European culture to the north of the Alps characterized by the first appearance of iron swords, exported weapons of high quality to several European countries, among them Greece. Toward the end of the sixth century BCE, shorter swords and daggers began to appear.[20]

Iron swords became common in large regions of Europe during the Celtic expansion in the fifth century BCE. As the Celts—of which the Gauls may have been the most extensive population with their center of power "north of the Alps in the region now within the borders of France and Belgium and part of Spain"[21]—began to dominate Europe, their swords lent their shapes to the swords of the Greek and Roman empires. Note that one of the distinguishing features of early Greek society is the social strata, the division between free and slave. Many manufacturing jobs in Classical Greece, including blacksmithing, involved slave labor. For example, "the father of the [Greek] politician Demosthenes [384-322 BCE] owned two manufactories, one making swords with over thirty slaves."[22] The orator Lysias from the fifth century BCE recorded having 120 slaves in his family's possession, the majority of whom were employed making shields.[23] Iron ore was not only plentiful, "but the other needed ingredient to smelt it—charcoal—stood in rich supply in the forests that blanketed Europe."[24]

The early Greek swords used by infantry had straight two-edged blades, without taper, of equal width from hilt to point. In the fourth century BCE, the length of the sword was increased by order of Iphicrates, an Athenian general. This decision was perhaps a result of a military tactician finding the sword too short, since the length of the spear was increased around the same time.[25] Iphicrates made several other improvements to the armor worn at the time, and became known for his knowledge of military tactics. His many inventions allowed the soldiers to go forth under diminished weight and become more mobile, while simultaneously protecting their bodies through the better armor.[26]

Despite the increased length of the Greek infantry sword, the Spartan sword measured only 14 to 15 inches in length, and was by some considered so short and non-frightening "that a juggler would think nothing of swallowing it." Those who had an intricate understanding of edged weaponry and combat, however, knew that while a long sword carried certain psychological powers, length did not necessarily communicate the deadliness of the weapon. A sword of 14-inch length is plenty capable of piercing and killing a combat adversary. The Spartans were also pragmatic with respect to their weaponry and asserted that if a sword appeared a bit short, one had only to advance a pace in order to do the necessary damage.[27] When an Athenian mocked the short Lakedaimonian swords, a certain King Agis supposedly replied: "And even so we still reach our enemies with these swords."[28] This statement demonstrates that tactics and military prowess were valued more than the length or size of the weapon.[29]

Bronze figure of naked Spartan warrior with short sword, from the sixth century BCE. Image source: Davide Ferro, reproduced under Wikimedia Commons license.

The different sword shapes of the Classical period came with certain factual flaws, however. For example, a weakness in the blade near the hilt could cause the sword to break during repeated blows; a misfortune possibly suffered by the Lakedaimonians during the battle of Thermopylae in 480 BCE, as described by the great Greek historian Herodotus:

"[They] fought with their swords when their spears had broken, and then 'with swords (machairai), *those who still had them* [author's emphasis], and even with their hands and teeth.'"[30] The Romans, however, asserted that it was the Celtic iron that displayed the greatest signs of inferiority and tended to bend at first strike. This impression might have rested with the fact that swords were sometimes abused and used for tasks for which they were not initially intended. For example, the Nervii, Belgic tribes and enemies of Rome living in northern Gaul, having no supply of iron tools, were observed to use their swords as utility tools for digging their winter quarters, cutting the turf and emptying it out by hand and with their cloaks.[31] Note that conflicting sources exist. Although great variations in craftsmanship existed, some of the Celtic swords were said to be so well crafted that the quality of the blades could easily pass as steel.[32]

Celtic sword finds display handles made of bronze, blades of iron, and sword sheaths of bronze. Iron sheaths have also been found, although more rarely than bronze. Sheaths frequently had a loop through which a suspension chain of iron or bronze could be passed. Sword lengths varied from 1 foot 8 inches to 3 feet 6 inches. Although bronze handles have been found, handles were often made of wood which has not survived. Tangs were of moderate length and would fit into a handle of good size.[33] The Celtic Bronze Age swords were normally leaf-shaped. Those of the later Bronze Age had slightly tapering edges that were nearly parallel all the way to the point.[34] "Swords from around 400 BCE to the end of the Celtic Iron Age were long, two-edged, with straight parallel blades."[35] Note that bronze instead of iron was often used in the construction of armor, due to its greater flexibility and ease of ornamentation. Roman equipment discovered in southern Scotland has revealed that the soldiers wore arm guards made of thin bronze strips riveted onto leather straps, which overlapped upward in order to protect against a sword thrust from below.[36]

The weapons that have survived into present day provide a good source of information about the Celtic warriors. For example, finds that have been unearthed from the time that the Celts sacked the Etruscan city of Marzabotto, c. 350 BCE, include swords made of iron.[37] Since the sword displayed prestige, Celtic scabbards were often highly "decorated with precious metals and stones." Swords might also display hilts of gold or ivory. A sword from Kirkburn, Yorkshire has inlays of scarlet enamel on the hilt and scabbard. A reconstruction of this sword demonstrates great craftsmanship. The original sword "was made from over 70 pieces of iron, bronze, and copper."[38] The Celtic warriors also carried spear and shield as part of their panoply. Those who could afford it would add a helmet, iron breastplates, or chainmail. The goal of the Celts was to get to close quarters and fight aggressively with spear and sword, thereby proving their battle prowess. Throwing javelins from a distance, for example, would have been considered an act fit for cowards.[39]

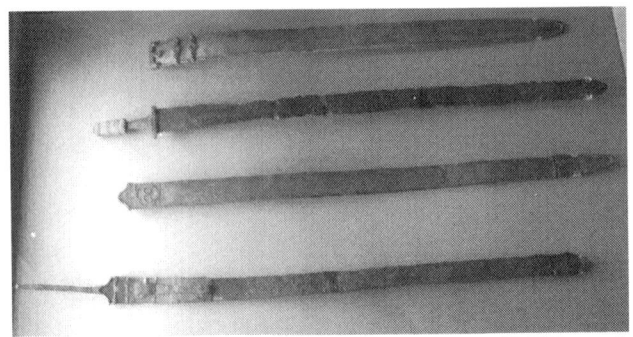

Celtic swords from the second century BCE to the first century CE. Celtic swords displayed prestige. Hilts and scabbards, in particular, were known to be richly ornamented with inlays of precious metals and stones Image source: Völkerwanderer, reproduced under Wikimedia Commons license.

By the time of Julius Caesar, the Celts dominated large parts of Britain; although, the Romans had begun to conquer Celtic lands in the third century BCE and would continue to do so until most of Britain would fall under Roman rule.[40] The Celts thus came into military conflicts with the Mediterranean world and ended up in numerous clashes with the Romans, some of which involved small numbers of combatants, others several tens of thousands. The Celts, who were less organized than their Roman adversaries, did not have as elaborate battle formations but relied instead on personal aggression, individual prowess, and a belief in psychological warfare, or the idea that they could frighten the enemy into flight or overwhelm him by brute force rather than elaborate tactics. If they managed to break a Roman formation through their wild charges, the Roman legionaries might find themselves outclassed by the big Celtic warriors whose slashing swords had greater reach than the Roman thrusting swords. Despite the onslaught, the Romans mocked these "barbarian" forces, accusing them of lacking endurance to fight beyond the first assault.[41] Unable to engage in pitched hand-to-hand battle, the confidence of the Celts began to erode, making them easy targets for Caesar's armies.[42]

While the Romans attributed their successes to the discipline of their forces—it was not the sword *per se* that gave the Romans their tactical advantages and allowed them to conquer a large part of the world, but their discipline and cohesion on the battlefield, and sometimes their leadership; for example, Julius Caesar motivated or forced the soldiers to do his bidding and not waver under attack—nineteenth century British explorer and soldier Richard Burton disagrees and credits Roman victories, particularly in the Second Punic War and the conquest of Gaul, to the superior weaponry of the Roman army against an enemy who still relied on swords made of bronze. Although the Gauls were essentially swordsmen with a history that included weapons of stone, copper, bronze, and iron, the Gallic sword had a blade that was "mostly two-edged, about one metre long, thin, straight, and without a point; it had a tang for the attachment of the grip, but no guard or defence for the hand." Its long and thin shape made it inferior to the shorter and stronger Roman swords. The Gauls wore their sword hanging to the right. The Gallic auxiliaries of Rome wore the sword on the left. The difference might be that the Gauls also carried a dagger.[43]

The Roman swords, by contrast, were of the Greek leaf-shape but larger and heavier than their Greek counterparts and of varied length between 19 and 26 inches. The gladius, which may be the type of sword most frequently associated with the Roman soldier, measured around 22 inches with a 6-inch grip and sometimes a crossbar of around 4 inches in length. This type of sword had straight edges narrowing slightly toward the point, and was considered effective for heavy fighting at close combat range. When the Romans invaded the Spanish peninsula in 219 BCE for the purpose of subverting the Carthaginians, they found the Spanish sword highly capable and adopted its shape for their own future sword production. After the battle of Cannae, steel rather than bronze was used more universally, leading to stronger blades and a greater ability to conquer enemy forces. Iron swords, although probably manufactured in large quantities, have rarely been found. A reason might be because iron proved highly corrosive and might not have outlasted the tooth of time.[44]

The bronze hilt with a simple cross guard, a wooden grip, and a metal pommel often in the shape of a pyramid and sometimes decorated was retained for some time. The idea of decorating the pommel was purely practical and did not indicate nobility status, as was the case, for example, in medieval Japan and Europe. Rather, the decorations were thought to increase the value of the sword to the individual soldier, so that he would be less reluctant to lose it on the field of battle or part with it in flight. The sheath was made of wood or leather with a reinforced metal tip. The Roman soldier also carried a two-edged stiletto-type dagger of copper, bronze, or steel as a sidearm to the sword. The sword is thought to have been worn on the right side, although some confusion exists indicating that it may have been worn on either side as the soldier saw fit.[45]

In the early Classical period, the straight sword was preferred among infantry forces over the curved sword which was considered more practical for men fighting from horseback or chariots, mainly the eastern peoples of Turkey, Egypt, and Assyria. The slash of the curved sword seemed more anatomically correct for cavalry warfare, where the natural motion of the wielder's arm allowed him to use the momentum of the horse to his advantage. Because of its thrusting characteristics, the straight sword, by contrast, was difficult to handle from horseback with accuracy particularly when galloping at full speed toward the enemy. While curved swords were also used in infantry armies, straight swords have less often found use in cavalry. Since the thrust generally proved deadlier than the cut and could be delivered without a great deal of strength, it benefited men of smaller build who did not have the momentum of the horse to their advantage. According to Richard Burton, the Romans were of light build and possessed less muscular strength than many of their eastern neighbors, which is a reason why they preferred the straight over the curved sword. Note, however, that Burton also states, to the contrary, that the people of the Roman Empire were admired for their great physical build and strength, and that the "Roman soldier generally prevailed against races whom he excelled in size, weight, and muscular strength."[46]

Additionally, favoring the thrust over the cut, the Romans held the view that those who cut with the sword were easy to conquer (a reason why they felt their Gallic enemies were inferior). While a stab to a vital organ almost always proved deadly, a cut with the edge, even if made with significant force, would seldom result in an immediate kill. Moreover, the vital parts of the body were often protected by armor and, naturally, by bone, and the thrust proved more efficient for penetrating these vital areas. Often only two inches penetration was needed to produce a fatal wound. An added benefit was that while a slashing attack would expose the right side of the body and thus jeopardize the soldier's safety against a countercut, the thrust was quicker and allowed the soldier to beat his opponent to the blow, because the enemy must receive the point of the sword before perceiving an opening for a countercut. Slashes were still used to target the unprotected legs of a combat adversary, however.[47]

Since a cut required significant room to execute, the Gallic warriors were limited in their ability to approach their enemy in formations as tight as those of the Romans, where swinging a long weapon would have proven detrimental to one's comrades. Greek historian and critic Dionysius of Halicarnassus (c. 60-7 BCE) observed that the Romans could duck under the arms of the Gauls as they raised their swords aloft to cut, take the cut on their shields, and counter with a straight thrust to the enemy's groin and into their vital organs.[48] Although the Romans borrowed sword technology from neighboring cultures including the Greeks, Spanish, and Celts, they improved upon the swords to fit their own idea of combat. The short sword of the Roman infantry soldier also promoted his ability to move the shield as necessary for protection.

The *spatha*, a straight double edged sword of 25 to 35 inches in length, longer than the typical gladius and with a slight taper into a rounded point, was an auxiliary weapon used by the Roman cavalry. It took the place of the gladius in the first century CE and was used until around 600 CE. Unlike the short infantry sword, the spatha was slimmer and used primarily as a slashing weapon.[49] Using his spear as his primary arm when on horseback and his spatha when on foot, the Roman cavalryman was a skilled fighter both from horseback and dismounted.[50] A Roman tombstone from the last decade of the first century CE, displayed at the Römisch-Germanisches Museum in Germany, depicts a cavalryman attacking a fallen enemy combatant with a spear, while his long spatha is hanging sheathed with a strap around his belt by his right side. The carrying position of the spatha on the right can be verified through engravings on several cavalry tomb stones. It has been speculated that the spatha was also worn on the left side on occasion from the second century CE.[51]

While the Romans adopted and refined the weaponry used by neighboring regions, the native people of Spain, called the Iberians, were likewise influenced by the Classical world through extensive trade and the provisioning of mercenary soldiers to the countries around the Mediterranean Sea as early as the fifth century BCE. During the Punic Wars, Spain was a strong military force in the region and provided Hannibal Barca with some of its best troops. The Iberian sword can be divided into the *falcata* and the *espasa*, and had, according to Polybius, a point that was as effective for wounding an enemy combatant as was the edge.[52] Note that the history of the sword in Spain reaches several thousand years. The falcata, meaning sickle-shaped sword, forged by Spanish blacksmiths by the fifth century BCE evolved from the Egyptian kopis or khopesh. This sword, considered superior in quality, was adopted by Alexander the Great during his conquest of the world. With its curved inner cutting edge and widening blade toward the tip, the falcata proved a powerful chopping weapon. The espasa was more readily adopted by the Romans and resembled their two-edged straight short sword, also called the gladius hispanicus, or Spanish sword. When the Romans finally conquered the Spanish peninsula, it was with swords of Spanish origin.[53] The Iberian warriors of the fourth and third centuries BCE, like their Greek and Roman counterparts, carried a short stabbing sword in addition to a throwing and a thrusting spear, a round wooden shield, and a protective leather helmet and cuirass. Their panoply of weapons and armor allowed them to wage combat at both long and short range. In the late third century BCE, a bronze helmet was added along with longer swords that had both thrusting and slashing capabilities.[54]

Around this time, Iberia became an "important logistic base for Carthage in its struggle with Rome, providing silver, raw materials for the war effort, and soldiers. The peninsula quickly became a battlefield when the Romans landed in Ampurias in 218 [BCE], trying to sever Hannibal's supply lines." The traditional Iberian weaponry was modified to suit the new high-intensity combat situation, with the short and straight sword becoming the most important offensive weapon and taking precedence over the earlier combination of spear and sword.[55] The Spanish sword was considered so formidable, according to Titus Livius, that it had the capacity to chop bodies to pieces, tear arms away at the shoulders, and separate heads from bodies while laying all the vitals open. The Spanish sword might have been responsible for more deaths than any other weapon in Classical times.[56]

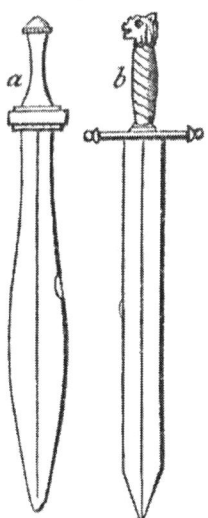

(*a*) Greek and
(*b*) Roman
Swords.

Greek and Roman sword comparison. Image source: Rev. Thomas Davidson, Chambers's Twentieth Century Dictionary of the English Language, reproduced under Wikimedia Commons license.

SWORD TRAINING

The sword to the Greek forces was considered mainly an instrument of last resort used in close combat on the field of battle, after the spear had been thrown. It was used in a hacking manner, and despite the crudeness the Greek soldier displayed a great degree of finesse with the weapon. Knives and swords were handled from childhood. Although agility proved important, a big and strong soldier was valued more than a man of slighter build.[57]

The Romans adopted many of their weapons, training methods, and culture from the Greeks. The gladiatorial spectacles, however, were perhaps the greatest displays of combat aptitude and reinforced the importance of training. The gladiator swords and spears, normally of high quality, were made either of steel or tipped with steel. Both straight double edged and curved swords existed, including the leaf-shaped Greek sword. The Thracian gladiators (who occupied the area of southeastern Europe: Bulgaria, northeastern Greece, and European Turkey), by contrast, favored a short knife. Although, by today's standards, the gladiatorial displays seem cruel, Richard Burton, in his writings, emphasizes the difference between cruelty and brutality. In Burton's view, the former encompasses "greatness of intellect," and the latter "debasement." Moreover, how one characterizes a sporting event or combat activity and whether it is considered "worthy," has a lot to do with whether or not it includes an element of fair play. Without fair play, it becomes merely an act of barbarism. According to Burton, the gladiatorial games of Rome were regarded as "proper" and courageous displays of combat skill.[58]

The Romans thus prioritized the study and use of arms, and encouraged young men to practice their skills. The training methods used to train gladiators were gradually adopted for use by the military. As previously noted, the Roman legionaries did not view the sword as a symbol of status and nobility, as did the Japanese samurai and the European medieval knights. Rather, it was a functional tool issued to the soldier by the army so that he could properly do the job he had been assigned. Sword training included driving a stake into the ground and beating it repeatedly with a wooden practice sword simultaneously using one's shield for defense as if the enemy were to counter the attack, thereby conditioning both body and mind for battle. When the recruit had achieved proficiency with the practice sword and displayed the mental attitude required to engage an enemy in combat, he was given a real sword.[59] According to Flavius Vegetius, whose writings date to around the fourth century CE and whose military treatises came to have a heavy influence on European battlefield tactics, the recruits were given "bucklers woven with willows, twice as heavy as those used on real service, and wooden swords double the weight of the common ones. They exercised them with these at the post both morning and afternoon." The heavier swords and bucklers used in training would hopefully benefit the fighter's strength and speed in a real engagement. The soldier struck a post about six feet in height that was firmly fixed in the ground, aiming his strikes at the head, sides, and legs of the training dummy with both the point and edge of the sword, while advancing and retreating as would be needed in the real fight, simultaneously maintaining awareness of the surroundings, ready to defend against an enemy combatant's counterattack.[60]

Training was also emphasized for the purpose of preventing mutiny in the army. It was reasoned that when the soldier was frequently called to roll and expected to be ready for drill and military service at a moment's notice, when he was tired after a full day of training, two things would occur: The skills he had acquired would inspire him to remain in the army, and he would be too exhausted by the day's end to consider mutiny.[61] During Rome's battles against Hannibal Barca, for example, the land forces were made to march in full armor on the first day of training, to polish and repair all their arms on the second day, to rest on the third day, and on the fourth day to practice sword fighting with wooden swords "covered with leather and with a button on the point, while others practised casting with javelins also having a button at the point. On the fifth day they were to begin the same course of exercise again."[62]

Training benefited not only the infantry soldier. Strength and agility were also needed to mount a horse in an instant; to vault onto the horse with equal dexterity from either side with drawn sword or lance in hand amidst the confusion and chaos of battle.[63] When the would-be cavalryman had learned how to ride and control the horse and its movements, he would begin to train with shield, sword, and other weapons he might be using, employing one stroke after another with the sword while in motion in the attempt to pursue and overtake the enemy. He would practice striking while riding alongside an enemy cavalryman, and would train in cross-country riding so that both horse and rider could get used to working together while undertaking the duties of war.[64]

Roman cavalry reenactor wearing a replica spatha. Image source: David Friel, reproduced under Wikimedia Commons license.

MILITARY ORGANIZATION AND BATTLEFIELD TACTICS

In ancient Greece, young men whose focus was the state were recruited to serve as citizen-soldiers and, upon passing their training, earned privileged positions in society. A Cretan drinking song expressed the importance of soldiering: "My wealth is spear and sword, and the stout shield which protects my flesh . . ."[65] Since the soldier likely had obligations to home and family, he could not afford to go on campaign for years at a time, however. A short and decisive battle was preferred over siege warfare that starved the enemy into submission. Sparta, by contrast, proved unique in that it had a standing army. Soldiers were selected at birth and military training began in the childhood years. The Spartans were warriors first. Trained to obey and endure until the end, they fought as professional soldiers, while Helots, a serf-like social class, worked the lands.

The Peloponnesian War (431-404 BCE), the twenty-seven year long conflict between Athens and Sparta, according to Thucydides—a Greek historian of the fifth century BCE, famous for having authored the History of the Peloponnesian War recounting the battles between Athens and Sparta—was the greatest struggle in the history of Greece, a belief he based on the examination of earlier wars: "And though men always judge the present war wherein they live to be greatest, and when it is past, admire more those that were before it, yet if they consider of this war by the acts done in the same, it will manifest itself to be greater than any of those before mentioned."[66] The foundation for the conflict was laid through a variety of events. Thieves were crossing over from one island to another. Cities were being built in high country to avoid pirates and, since "houses were unfenced and traveling was unsafe, they [the citizens] accustomed themselves, like the barbarians, to the ordinary wearing of their armour." Moreover, "[t]he Corinthians quarreled the Athenians for besieging Potidaea . . . The Athenians quarreled the Peloponnesians for causing their confederate and tributary city to revolt, and for that they had come thither and openly fought against them in the behalf of Potidaea."[67]

The lengthy speeches in Thucydides' account of the war display good insight and advice about the various challenges: "Consider before you enter how unexpected the chances of war be. For a long war for the most part endeth in calamity from which we are equally far off . . . And men, when they go to war, use many times to fall first to action . . . and when they have taken harm, then they fall to reasoning."[68] Sparta, consisting of sparsely decorated scattered villages, seemed inferior to Athens and gave the illusion that Athens' power was "double to what it is." No man, however, "comes to execute a thing with the same confidence he premeditates it. For we deliver opinions in safety, whereas in the action itself we fail through fear."[69] Plutarch's *Moralia* mentions the use of short swords by the Lakedaimonians. "In answer to the man who sought to know why the Spartans used short daggers in war," the statesman and general Altalkidas answered, "Because we fight close to the enemy."[70] While the early Greek swords were designed for both cut and thrust in hand-to-hand fighting with a single enemy combatant, the close order of battle that had been drilled into the Lakedaimonians from early age made the short sword a necessity. Designed primarily as a thrusting weapon to be used in underhand fashion against the trunk of an enemy, it was employed after the soldier had pushed through the enemy's shield wall; a task for which one could use one's shield, and which would throw the enemy off balance and create the space needed for wielding the sword.[71]

The Greek hoplite, by contrast, carried the sword as a sidearm to the spear which he relied upon for piercing his enemy's shield or any body part unprotected by armor. The spear could easily shatter during the initial collision of the forces, however, or break when striking the breastplate of an enemy's armor.[72] When he lost his spear, the hoplite had little choice but resort to the sword, often slashing at his opponent's legs, in hand-to-hand combat.[73] If a hoplite fell to the ground in battle, perhaps as a result of a blow to the legs, he might attempt an upward thrust with the sword into a spear wielding enemy combatant with the hope that he would at least inflict some damage, if not a killing blow, before he was himself killed.[74] The soldiers were successful not because of the equipment they used, but because of the phalanx formation composed of a large group of men designed to function in coordinated action.[75] Force was established early in the encounter, with the enemy driven back or into flight so that hand-to-hand combat could ideally be avoided. When the enemy was another phalanx rushing forward with equal strength, the fight deteriorated quickly until both sides "were butchering each other with their secondary weapons." In the melee, space limitations might deprive the hoplite of the use of both spear and sword. He was then forced to resort to his bodyweight to push the enemy to the rear. The hard bronze armor he wore would protect him against an enemy's weapons and, in best case, allow him to dominate the fight.[76] The shield proved more important than the sword because it also protected one's comrades, allowing each man to contribute to the cohesion and momentum of the phalanx.

The Greeks also used edged weapons in silent killings in siegecraft for the purpose of disposing of sentries. For example, the Plataeans chose to "make a sortie against the wall of the Peloponnesians" and put the sentries to the sword. (Plataea was an ancient Greek city which was destroyed in the Peloponnesian War. The wall of the Peloponnesians was constructed for the purpose of defending against the Plataeans and the Athenians.)

> Such being the structure of the wall by which the Plataeans were blockaded, when their preparations were completed, they waited for a stormy night of wind and rain and without any moon . . . Crossing first the ditch that ran round the town, they next gained the wall of the enemy unperceived by the sentinels, who did not see them in the darkness, or hear them, as the wind drowned with its roar the noise of their approach . . . those who carried the ladders went first and planted them; next twelve light-armed soldiers with

only a dagger and a breastplate mounted . . . Meanwhile the first of the scaling party that had got up, after carrying both the towers and putting the sentinels to the sword, posted themselves inside to prevent any one coming through against them . . ."[77]

The wars against Persia, a series of conflicts fought between several of the Greek city-states and the Persian Empire in the fifth century BCE, are regarded as defining wars of ancient Greek history. According to Herodotus, the Persians were deficient in armor and inferior in skill, but were still as courageous as their adversaries: "The Persians fight bravely hand to hand until they are demoralized by the death of their commander."[78] Note that in Roman eyes, the Greeks were lazy and unorganized. No matter how many arms they were provided or how often they went to war, they were "over-educated, luxurious, soft . . . and scared of their own shouting."[79] The Carthaginians did not enjoy a much better reputation. Despite the military feats of their great general Hannibal Barca, they were considered treacherous, cruel, superstitious, and effeminate. If, by chance, they managed to score a victory, it was because they "had relied on the courage of others to fight their battles."[80] Simultaneously, both Rome and Greece believed that their type of warfare was characterized by courage in hand-to-hand fighting among infantry soldiers rather than cavalry, and that discipline was important to securing the victory.[81]

The famous 300 who defended the pass at Thermopylae against the great Persian army "hewed down [with their swords] the ranks of the Persians." Many warriors, both Persians and Spartans fell in battle that day, including the Spartan chief Leonidas. At the "entrance of the straits," a stone lion was erected in his honor:

> Here they defended themselves to the last, such as still had swords using them, and the others resisting with their hands and teeth; till the barbarians, who in part had pulled down the wall and attacked them in front, in part had gone round and now encircled them upon every side, overwhelmed and buried the remnant which was left beneath showers of missile weapons.[82]

Diodorus Siculus, a Greek historian of the first century BCE, noted how the Grecians and the "barbarians" were both reported as having fought with "great obstinacy . . . without regard or fear of death." In their strife for glory, the Athenian and Lakedaimonian generals strove to exceed each other in military valor. When the Persian camp was finally taken, no quarter was given to the enemy for fear that their greater number might regain their strength and attempt some "mischief." More than a hundred thousand Persians were thus "put to the sword," after which the Grecians proceeded to bury their own ten-thousand dead.[83]

Although the foot soldier was hailed more than the cavalryman in ancient Greece, cavalry did exist. The Greek cavalryman of the fifth and fourth centuries BCE preferred the javelin as his primary weapon, sometimes carried in pairs, one to be thrown and the other to be used as a thrusting weapon in either overhand or underhand fashion. He carried a sword worn at the waist as a sidearm. The cavalryman must display a willingness to close with his enemy and fight with the sword. The Macedonian cavalry of Alexander the Great was likely trained for close combat with a lance of significant length (seven to ten feet). Once the cavalryman had scored a strike against his enemy or enemy's horse, he would drop his lance and take to the sword. Under less favorable circumstances, cavalry clashes between individual soldiers could take the form of closing with drawn swords. The enemy combatants would then drop the reins and grab each other while falling to the ground and battling it out with the sword.[84]

Anabasis, Xenophon's account of his march to the Black Sea, compares the strength of cavalry and the foot soldier, suggesting that a force of ten thousand horses strong can still carry only ten thousand men on their backs. (Xenophon who had extensive experience dealing with the Persians differed from the traditional Greek warrior in that he preferred a saber-type sword to the straight blade, because of the ease with which a cutting stroke could be delivered.) According to Xenophon, the foot soldier was "well planted upon the earth" and could defend against heavy blows easier than a mounted soldier, who was always at risk of being unseated from his mount. The only strength cavalry had over the foot soldier was the ability to take to flight quickly.[85] It was thus not the horse that proved dangerous, but the man who commanded it and the fighting skill and fortitude he displayed. Cavalry was often associated with the "effeminate" eastern warriors. A decoration on an ancient Greek jug used for mixing wine and water depicts a naked and muscled western foot soldier facing a softer looking eastern cavalry rider in battle.[86] Likewise, the lance and sword were not superior weapons *per se*. Rather, it was the skill of the soldier using the weapons that made him seem formidable and superior. Fighting skill was also increased in part because of the great amount of fighting that took place, which made the soldier more aware of the need to train to perfection. As military historian John Lynn points out, however, although the hoplites achieved decisive battle, it is important to recognize that battle was decisive because "both sides agreed [beforehand] that it would be."[87]

While the Greek foot soldier relied on the spear as the primary weapon, the Roman soldier replaced the spear with the javelin, which could be thrown from a distance in order to break up enemy formations and cause chaos in their ranks.[88] Swords were drawn after "shifting the javelins to the left hand," while rushing toward the enemy.[89] Vegetius made an interesting observation with respect to power when wielding the javelin versus the sword (for a right-handed fighter):

> It must be observed that when the soldiers engage with the javelin, the left foot should be advanced, for, by this attitude the force required to throw it is considerably increased. On the contrary, when they are close enough

to use their piles and swords, the right foot should be advanced, so that the body may present less aim to the enemy, and the right arm be nearer and in a more advantageous position for striking.[90]

Historical research suggests that the Roman soldier continued to use the javelin throughout the duration of battle. As the clashing lines of warriors separated during a lull in the fight, new opportunities for its use would emerge. The repeated clashes and lulls tended to make battles last for several hours. Battles of short duration were normally due to the weaker side realizing its inability to defend against the Roman onslaught and deciding to take to flight before the actual battle could commence. One might also consider the physical strength and endurance a warrior at the front needed in order to fight at close range with sword and shield, pressing against his enemy with no opportunity to take a breather. It is unlikely that a soldier could have sustained physical battle with the sword for several hours without letup, trusting only in his strength and fighting spirit until reserves could be brought up from the rear lines.[91]

As previously noted, the Romans also carried as part of their military equipment a short dagger as a sidearm, to be used for many tasks other than hand-to-hand combat. The Roman dagger or *pugio* (of Spanish origin) was commonly made of bronze or iron. Due to its two cutting edges, sometimes broadening toward the tip and then tapering to a point, it was primarily a stabbing weapon. Pugio and scabbard, richly ornamented with intricate silver and brass inlays, were worn by high ranking military men as a badge of distinction.[92] Daggers were also carried concealed and used as weapons of assassination. A striking example is the assassination of Julius Caesar in the Senate in 44 BCE.[93] The Romans wore the dagger on the right side of the body. The ancient Greeks, by contrast, when wearing a dagger as a sidearm, wore it suspended by the sword on the left side of the body.[94]

Reconstruction of a Roman centurion with cingulum (belt), pugio (dagger) and gladius (sword). Image source: Wolfgang Sauber, reproduced under Wikimedia Commons license.

The Legion, a citizen-militia composed of "heavy-armed infantry" preceded by light infantry armed with javelins and swords "who cleared the way for action," was the cornerstone of the Roman military organization.[95] The legions were divided into four classes, each distinct in age and equipment: First the youngest and poorest *velites*, then next to them the *hastate*, then those who were in the prime of their life, the *principles* (or principes), and lastly the oldest known as the *triarii*. "They divide them so that the senior men known as triarii number six hundred, the principes twelve hundred, the hastati twelve hundred, the rest, consisting of the youngest, being velites." The youngest soldiers carried sword, javelins, and a round shield about three feet in diameter. The hastati carried, besides their shield, a Spanish sword "hanging on the right thigh." This sword had two edges that could cut efficiently, and proved excellent for thrusting because the blade was strong and firm. The hastati also carried two javelins, a brass helmet, and lower body armor to defend the legs. "The principes and triarii [were] armed in the same manner except that instead of the javelins the triarii carr[ied] long spears."[96]

On account of their status as citizen-soldiers (both farmers and soldiers as needed), when called upon, the Romans "laid aside [their] tools and put on the sword." The farmers lived a tough life, which was a reason why it proved beneficial to recruit the soldiers from the country, "[f]or it is certain that the less a man is acquainted with the sweets of life, the less reason he has to be afraid of death."[97] The Roman soldier, a disciplined man in the possession of individual and national dignity, had a will so forceful and "persistent that it levelled every obstacle." Rome treasured the greatness that came with expansion. It was said that the only command the Roman soldier understood was *Forward!* Retreat, of course, was impossible. A standstill, considered the beginning of a retreat, was not looked upon kindly. Despite these individual qualities and the warrior-ethos drilled into the soldier, Rome suffered its shares of defeats and humiliation at the hands of Hannibal Barca.[98]

Any negative experiences, however, were thought to make the Roman soldier even stronger and would open the path to future victories. Each man individually or in common with his comrades could form a front in any direction to face any danger on the field of battle.[99] To remain effective, the soldier must not only have enough space to move separately from his peers in order that he could turn to meet a threat and cover himself with the shield, he must also be able to use the sword in both cutting and thrusting fashion. The Roman soldier occupied with his weapons "a space of three feet in breadth." Polybius believed that a looser order was required, however, where each man should be at a distance at least three feet from the man next to him, as well as those in front of and behind him.[100] As formidable fighters, confident that their comrades, shields, and swords which were strong enough to endure repeated blows, would protect them, the Roman lines proved difficult to break through.

Discipline in the legions was strict. Infractions such as stealing or lying resulted in heavy fines, including punishment by flogging. Roman soldiers were also punished for failing to give the battle the respect it deserved. For example, Tacitus, a senator and historian of the Roman Empire, describes how a man was executed for putting his sword aside while digging a ditch.[101] Furthermore, false boasting in order to gain distinction was considered unmanly and disgraceful, as was leaving or throwing away one's arms as a result of fear:

> Therefore the men in covering forces often face certain death, refusing to leave their ranks even when vastly outnumbered, owing to dread of the punishment they would meet with; and again in the battle men who have lost a shield or sword or any other arm often throw themselves into the midst of the enemy, hoping either to recover the lost object or to escape by death from inevitable disgrace and the taunts of their relations.[102]

Critognatus, a Gallic noble during the time of Julius Caesar, spoke about the shame of surrender, even as supplies were running short: "To be unable to bear privation for a short time is disgraceful cowardice, not true valour. Those who voluntarily offer themselves to death are more easily found than those who would calmly endure distress . . . What courage do you think would our relatives and friends have, if eighty thousand men were butchered in one spot."[103] After Caesar had pondered his enemy's speech, he drew the conclusion that men prefer to fight to the death over enduring privation, even for a short time, and exclaimed that "it was better to be slain in battle than not to recover [our] ancient glory in war."[104] Although Caesar gave his men command to act, he "censured the rashness and avarice of his soldiers" and accused them of being arrogant, "because they thought that they knew more than their general concerning victory, and the issue of actions: and . . . he required in his soldiers forbearance and self-command, not less than valour and magnanimity."[105]

In an act intended to encourage his men to pursue combat as a unit and discourage them from taking to flight, Caesar removed all the horses, his own included, then had the soldiers break up the enemy phalanx by attacking from higher ground, hurling javelins first before making a charge with drawn swords toward the opposing force. This tactic proved successful because the bucklers that the Gauls were using to defend themselves failed to hold up under the onslaught of javelins, and were quickly pierced through and otherwise damaged. Many of the javelin points stuck in the shields, causing a hindrance to the mobility of the Gallic troops. As they continued to take heavy casualties, they were finally forced to give ground to Caesar's army.[106]

A striking example of the difficulty a javelin could cause the defender, even if it failed to kill, is related by Caesar when Titus Pulfio, one of the Roman centurions fighting the Nervii, had his shield pierced by a javelin which then stuck fast in his belt, a "circumstance [which] turn[ed] aside his scabbard and obstruct[ed] his right hand when attempting to draw his sword." Luckily, Lucius Varenus, Pulfio's rival and competitor but also his safeguard, drew attention away from the incident. Together they managed to disperse the enemy troops, bring in additional relief forces, and save their lives while retreating to fortifications.[107] Although the preferred tactic was to hurl javelins from a distance, it could also happen that the enemy would rush forward so quickly that the Romans were forced to throw their javelins aside and fight hand-to-hand with swords from the outset. Securing victory would now prove more difficult, unless reinforcements were called upon to relieve the distressed soldiers.[108]

Swords were also used as psychological weapons to instill fear, discourage further attack, or send the enemy to flight (much like the bayonet has been used in modern warfare). For example, after a failed assassination attempt plotted against Comius, a king of the Belgic nation believed to have raised a conspiracy against the dictator but later becoming an ally and envoy of Caesar, swords were drawn on both sides more for the purpose of avoiding a fight than continuing with an all out melee. Escapes could now take place with the hope that Comius would die from the severe wound he had received in the head, or that the Gauls would learn their lesson and not come within sight of any Roman again. Likewise, when all the javelins had been thrown and the Romans charged forward with their swords, they frequently persuaded the enemy, after having killed a few, to take to flight rather than stand and fight hand-to-hand. The Roman army did not go without loss, however. In a battle against Lucius Afranius, one of Pompey's lieutenants, Caesar describes "the various turns of fortune," the hardships they suffered and how the legionaries were reduced by the swords of the enemy. Yet, despite the disadvantages his men suffered by fighting from lower ground in inferior numbers, they maintained the battle for five hours and drove the enemy into town. Approximately 70 of Caesar's men including a centurion fell in the battle, and 600 were wounded. The opposing party suffered five dead centurions and more than 200 other casualties.[109]

When counting war casualties, Caesar generally depicts the enemy losses as quite severe. His writings were probably affected by personal bias and the intention to appear powerful by portraying his own casualties as less significant. One account portrays the enemy surrendering and sending ambassadors to Caesar. When recounting the calamity of their state, they learned that the enemy senators "were reduced from 600 to three; that from 60,000 men they were reduced to scarcely 500 who could bear arms." Caesar does balance the account to some degree, however, by giving the enemy credit for their courage: "When the foremost of them had fallen, the next stood upon them . . . and fought from their bodies," and those who survived "cast their weapons against our men, as from a mound."[110] Caesar (according to himself) also considered it a general's duty to conquer as much by tactics as by the sword, and would rather have accomplished his mission without the risk of injury or death to his soldiers.[111]

When the fighting spirit was such that neither army would retreat and neither could be overpowered by force, a cavalry charge could possibly stir up some confusion in the opposing army and cause disorder in their foremost battalions. If this tactic failed to open a pass to the enemy, the Roman soldier would consider it a task for the infantry: "[C]ome on, as ye shall see me making way with my sword, in whatever direction I shall advance into the enemy's line, so let each man, with all his might, beat down those who oppose him."[112] Cavalry, in conjunction with infantry, was also responsible for policing actions guarding against public disorders. In one such event, as described by Herodian, a Roman civil servant, fully armed cavalry charged and cut down with the sword everyone in their path, and trampled with their horses those who had fallen.[113] To control the movements of the horse, the Roman cavalryman, due to the need to hold the shield and wield the weapons confidently with both hands, relied on his legs rather than a bridle.[114]

Tombstones of Roman cavalry with spear and sword. Note the fallen enemy trampled by the horse. Image source: C. Iulii Caesaris belli gallic, reproduced under Wikimedia Commons license.

Much of Caesar's political life was devoted to the conquest of Gaul, or the area that today is Western Europe, mainly France. Although the Romans criticized the Gauls and their weapons as inferior, as previously noted, many Gallic swords were of good quality. Moreover, the Gallic warriors were generally taller than their Roman counterparts. Their build in conjunction with their long swords most certainly instilled a feeling of respect, if not fear, in the opposing forces. The failure of the Gallic troops to fight in tight and protective formations like the Romans, however, left them vulnerable to the thrusting attacks of the Roman short sword and complicated both attack and orderly withdrawal. Once the battle came to close quarters, the space that the Gallic soldier needed for swinging his long sword or halting the Roman advance made it almost impossible to maintain a tight formation, even if this had been the desired tactic. Many of the Gallic warriors were also limited in their weapons and armor due to the expense of such equipment, and were therefore more vulnerable to a sword thrust than their Roman counterparts.[115]

The battle of Adrianople in 378 CE, perhaps considered the beginning of the end of the Roman Empire, ended in a devastating decisive defeat for the Roman army, where the "barbarians made Roman commanders look . . . utterly incompetent in the art of war." The Roman infantry had experienced some significant changes "[d]ue to reductions in available funds for military spending." The use of metal armor had become rarer and shields had become smaller. To compensate for the lack of defensive material, the short gladius, the trademark of the ancient Roman armies, was replaced by a longer sword. The previously proud Roman infantry, according to historian Ammianus Marcellinus, "became so huddled together that hardly anyone could pull out his sword, or draw back his arm." The Roman defeat at Adrianople was more a result of deficient tactics and strategy than of deficient weaponry, however. The Roman army suffered from insufficient numbers of high-quality troops to deal with the threat, as most troops "were committed to guarding imperial borders from Britannia to Syria." Moreover, the commanders acted with arrogance. In the belief that they were in charge of an army far superior to their barbarian enemy, they failed to ensure that the odds were stacked in their favor prior to entering battle.[116]

Following the loss, the Roman army ceased to be truly Roman. Increasing numbers of Goths came to fill the depleted Roman ranks. Cavalry attacking from long range, rather than infantry fighting up close with short swords, came to dominate battle. Perhaps the most important change was how one came to view the Roman soldier. Being Roman had meant having citizenship and had provided a legal framework for military service. By the mid fifth-century, the now mostly mounted Roman armies owed their allegiance not to the state as in the past, but to powerful warlords.[117]

WAR WOUNDS AND PROTECTIVE ARMOR

In 55 and 54 BCE, Caesar crossed the English Channel in order to prevent the Celts from launching an attack against the Roman Empire. The Celtic warriors, relying more heavily on cavalry than the Romans, had increased the length of their swords to almost 35 inches. Although considered too long to be wielded successfully by infantry, these swords enabled the Celts to ride into battle, dismount, and engage the enemy hand-to-hand. As observed and described by the Greek historian Dionysius of Halicarnassus (60-7 BCE), the Celtic warrior would raise his sword high above his head and allow it to drop onto the target with the full force of his weight behind the blow.[118] The sword suspension chain, developed in the third century BCE, allowed the soldier to twist his body "freely in fighting."[119] A good strike could produce enough power to cut through the enemy's shield and armor as well as bone, particularly if one followed up with several slashes to the body.[120]

There were obviously many ways in which a soldier could get wounded or killed. Sword wounds could be caused by strikes to the head, particularly if the combatant lacked a helmet. If the battlefield situation allowed, attackers might rely on feints to open up a target. For example, a feint to the opponent's head could make him raise his shield and expose his legs. (It is likely that the ideal situation would quickly deteriorate when entering the fray, however, forcing the soldier to assume a strict position of offense without a lot of consideration for any particular tactics beyond a hack and slash. What appeared as a deliberate feint and attack may merely have been a chance blow.) Any cut to the tendons at the back of the leg would result in a useless leg and would likely end the fight. Such a blow could also cause heavy bleeding and shock, particularly if the large blood vessels were severed. A cut to the hamstrings, which are responsible for leg movement as required when bending the knee, would likely cause a crippling injury. A cut to the leg required good timing, however. The combatant would have to drop his blade and then slice upwards, and would thus risk exposing his own back or shoulder to a countercut.[121] Greaves, or leg protection in the form of "thin sheets of bronze extending from the kneecap all the way down to the ankle," afforded the soldier some protection against sword cuts aimed at his lower legs where the downward movement of the shield failed to give adequate protection against a sword or spear thrust, or against missile weapons thrown from a distance.[122]

The western military tradition employed impressive armor as early as 600 BCE. The Greeks developed a variety of body armor such as helmets and shin greaves made primarily of bronze. The use of armor also reflected the value one placed on the soldier. He had undergone a significant amount of training to prepare him for war, and his life was not to be wasted. Moreover, the armor proved efficient for the phalanx movements and battlefield formations used by the Greeks in the early period.[123] A drawback was that armor proved cumbersome to the individual soldier when the phalanx broke and he could no longer rely on his comrades for protection. The heavily armed hoplites, for example, would typically cast aside whatever weapons they could to lighten their load and increase their chances of survival when fleeing cavalry or a lightly armed enemy. Thucydides remarked that an Athenian defeat in 413 BCE had resulted in "more arms . . . left behind than corpses."[124]

The estimated weight of Greek hoplite armor including sword, shield, helmet, breastplate, and greaves is estimated between 50 and 70 pounds.[125] When considering that the soldier had to exert himself extensively in combat, and often under the burning sun of the Mediterranean summer, it is less surprising that a man might choose to rid himself of as many weapons and pieces of armor as possible, particularly when it became evident that he could no longer win the battle. Even armor that was not heavy, such as greaves, would most certainly have proven cumbersome in a losing battle and would have hindered the soldier's movement.[126]

Combat in Classical Greece rarely resulted in total annihilation of the enemy forces, however. The Greek weapons and armor often gave sufficient protection against repeated attacks as the huge phalanxes, having made the initial clash, degenerated into "pushing contests" until either side could no longer oppose and chose to withdraw. Wounds to the chest which housed the vital organs would more often result in death than wounds to the extremities (although, these could cause infection and death). The bronze breastplate did at least a reasonably good job of protecting the soldier, and several successive piercing blows were normally needed in order to cause death. For example, "the fourth-century Spartan king Agesilaos was severely wounded, although not mortally so, only after he was caught surrounded and subjected to a multitude of spear and sword blows."[127] Each blow that failed to kill would give the soldier an opportunity to strike back at his opponent.

Horses were likewise susceptible to wounds. *The March Up Country*, the account of Xenophon's return from Persia where he had been aiding Cyrus the Younger in an attempt to take the throne from Cyrus' own brother Artaxerxes II, details the hostile territory, difficult terrain and weather, and the troops' encounters with enemy forces as Xenophon leads the Greek mercenaries, known as the Ten Thousand, on the long march to the Black Sea. Cyrus, according to Xenophon, was surrounded by "his bodyguard of cavalry about six hundred strong, all armed with corselets like Cyrus, and cuirasses and helmets." Moreover, all the horses "wore forehead-pieces and breast-pieces, and the troopers carried short Hellenic swords."[128] The vulnerability of the horses is also evident in Plutarch's account of the Life of Alexander the Great:

> While Alexander's cavalry were making such a dangerous and furious fight, the Macedonian phalanx crossed the river and the infantry forces on both sides engaged. The enemy, however, did not resist vigorously, nor for a long time, but fled in a rout, all except the Greek mercenaries. These made a stand at a certain eminence, and asked that Alexander should promise them quarter. But he,

influenced by anger more than by reason, charged foremost upon them and lost his horse, which was smitten through the ribs with a sword . . .[129]

Although all wounds did not result in death, any cut could easily cause infection and kill the solider or horse at a later time. Alexander received a sword wound to the thigh. To avoid giving his opponent credit, he stated in a letter to Antipater (regent of Macedonia and supporter of Philip II and Alexander the Great during the early years of Alexander's invasion of Asia) that it was merely a dagger that had wounded him.[130] He survived the injury and proved lucky once more during his encounters with the enemy. When attacking the Mallie, said to be a most brutal and warlike people, Alexander barely escaped with his life. He was at first thought dead, but recovered in his tent after an arrow head had been removed from his ribs:

> Then, as he brandished his arms, the Barbarians thought that a shape of gleaming fire played in front of his person. Therefore at first they scattered and fled; but when they saw that he was accompanied by only two of his guards, they ran upon him, and some tried to wound him by thrusting their swords and spears through his armour as he defended himself, while one, standing a little further off, shot an arrow at him with such accuracy and force that it cut its way through his

breastplate and fastened itself in his ribs at the breast. Such was the force of the blow that Alexander recoiled and sank to his knees, whereupon his assailant ran at him with drawn scimitar, while Peucestas and Limnaeus defended him. Both of them were wounded, and Limnaeus was killed; but Peucestas held out, and at last Alexander killed the Barbarian.[131]

Accidental self-inflicted wounds could also occur in the fray. Herodotus recorded the events that took place between the Greeks and the "barbarians," so that the "great and wonderful actions" would be preserved in all their glory, and noted how the "button" of the sword sheath fell off as Cambyses, son of Cyrus, sprung upon his horse, causing the point of the sword to enter his thigh.[132] (Note that modern soldiers often have reinforced tips on their knife sheaths to prevent the knife from causing an accidental injury.)

The ancient Romans, too, used body armor in the form of cuirasses made of iron hoops supported by a leather tunic. The Roman legionaries of the third century BCE wore an iron helmet which, in addition to providing good neck protection due to the strength of iron, was less expensive to manufacture than the Greek helmet made of more flexible bronze.[133] The Roman cavalry helmets had nape guards to protect the neck against sword blows, and cheek guards that came together at the front to protect the sides of the face and the throat. Scale armor made of small pieces of copper alloy or iron proved flexible and gave reasonably good protection against slashes; although, a good thrust with a sword or spear could still penetrate the armor. The Roman cavalryman must also be aware that armor did not provide sufficient protection against an upward thrust coming from infantry soldiers, which was why a new type of cuirass, not as flexible as scale but capable of providing better protection, was developed in the second century CE.[134]

Helmet from the Imperial Roman Army. Note the nape and cheek guards. Image source: Matthias Kabel, Museum Carnuntinum, reproduced under Wikimedia Commons license.

Grave finds from the time when the Romans invaded Britain reveal that extensive combat with the sword took place. Many of the skeletons that have been unearthed show signs of sword wounds. The Romans had to adapt their warfare to meet the onslaught of the Celtic warriors, for example, by adopting Celtic types of shields which were oblong, and by weakening "the Celtic charge with a volley of javelins and then use their shields to take the full weight of the Celtic slashing swords, while they stabbed at their enemies' guts," until the Romans finally turned the tables on the Celts.[135] The Roman gladius, however, despite its crudeness and relative simplicity, proved lethal. "[T]he Hellenistic Greeks were astounded at the carnage—severed limbs, decapitation, disembowelment—that Roman swordsmen could inflict on pike-wielding infantry with such a short sword."[136]

Roman soldiers also learned from Spanish soldiers who they had met on the field of battle, and adopted some of their techniques involving the point of the sword, instead of the edge, in a thrusting fashion. They would target their opponent's sides with the edge of the sword and the vital organs with the point, in addition to the tendons that controlled movement in the extremities.[137] Those who fought to defend the glory of Caesar (according to historical sources) displayed such courage in battle that if their sword hand were cut off, they might still strike the enemy in the face with the buckler held in the other hand.[138]

CONCLUDING REMARKS

Long range projectile weapons, primarily the throwing spear and javelin, were considered the weapons of choice in the Classical world. The sword was viewed as a sidearm to be used when one had ran out of javelins or lacked the manpower capacity to fight the enemy from long range. When Rome began to expand its empire into the Italian peninsula around the fourth century BCE, conflict naturally arose as the Samnites who occupied southern Italy disapproved of the Roman expansion and conquest of their territory. The swords of this period were essentially the same as the Greek leaf-shaped xiphos. With the Republican era that started to emerge around the second century BCE came a class system, in which military service was viewed as the province of the rich who were responsible for supplying their own arms. The Roman Empire continued its expansion into Spain, which led to the introduction of the gladius hispanicus, or the Spanish sword.[139]

Large numbers of soldiers were needed due to the continued drive for expansion, which also opened enlistment opportunities to ordinary citizens. Unlike the rich, however, these could not afford to supply their own arms and equipment. Standardization of weapons and armor therefore became necessary and included a bronze helmet, body armor, different projectile weapons, and the gladius, which took a straight double edged form and measured around 19 to 20 inches in length. The straight shape could be manufactured at a lower cost, which allowed the Empire to equip great numbers of soldiers. The gladius, which proved effective for stabbing through enemy armor and hacking at the enemy at close combat range, became the standard edged weapon sidearm for the Roman legions for much of the remainder of Rome's conquest and expansion. The need also arose for more mercenary and auxiliary troops, some of whom relied on their own weapons of native design.[140]

The gladius became a trademark of the Roman troops not because of the design, however—enemy forces frequently wielded swords of at least equal quality—but because of the discipline of the Roman legions. The Romans fought not for individual glory but for the benefit of the group. Cohesion in the lines gave them tremendous momentum without risk of splitting the formation and turning the battle into a free-for-all hand-to-hand fight without organization or plan. Unlike many other cultures—perhaps particularly the Japanese, but also the medieval Europeans—who revered their swords and viewed them as weapons fit only for the educated nobility, the Romans viewed their gladius merely as a practical battlefield sidearm. The Roman legionaries differed from other warriors because they placed teamwork ahead of glory. During their heyday, they were a well-oiled machine that appeared almost unstoppable in their conquest and hunger for greater territory.[141] A Roman bronze greave depicting a relief of Mars, the god of war, demonstrates the importance one placed on warfare: Mars wears "a Corinthian helmet, muscled cuirass, greaves and sword, and carri[es] a shield and spear."[142]

Gladiatorial combat might be a demonstration of the social development of the sword in ancient

Rome. It started as a "human" function to allow slaves and captives to fight for their lives rather than slaughtering them upon a pyre. Gladiators were also sent as volunteers for the purpose (or honor) of demonstrating their courage in front of the people. Others resolved disputes by the sword much like the later duels of honor in Europe; although, the near modern European version seldom resulted in the death of either combatant. The gladiatorial spectacles did not always proceed in an honorable fashion, however. Although most participants were well-trained and courageous, one story tells of a gladiator who fought in an amateurish and cowardly way by entering the fight heavily armored with a heavy sword, while his opponent fought only with flimsy blades of tin and lead.[143]

The Greeks, too, and particularly the Spartans, were notorious for their discipline and fighting spirit. They relied on a set of strict laws and had such cohesion in their phalanx that no warrior was considered superior or inferior to any of his comrades. Their training from early childhood, and their inherited piece of state land worked by Helots, allowed Spartan men to be full-time soldiers.[144] As expressed by Spartan general and poet Tyrtaeus, "Let each man hold, standing firm, both feet planted on the ground . . . let each man, closing with the enemy, fighting hand-to-hand with long spear or sword, wound and take him; and setting foot against foot, and resting shield against shield, crest against crest, helmet against helmet let him fight his man breast to breast, grasping the hilt of his sword or of his long spear."[145]

Unlike the regions of the Near East, which tended to rely on cavalry warfare, the ancient Greeks, although using the javelin initially to break up enemy formations, quickly closed on their foe and fought hand-to-hand with spear and sword.[146] Etiquette demanded that swords be left outside when entering the senate-house, however. Diodorus Siculus, referring to a law instigated by Charondas, a law giver from Sicily from around the sixth century BCE, notes how Charondas committed suicide when discovering that he had accidentally violated his own law.[147]

The sword in the Classical world was thus viewed primarily as a practical weapon to be used at close combat range in decisive infantry battle. It was carried as a sidearm to the spear and javelin and served less as a symbol of status and identity, than as a simple battlefield armament that aided the Greek and Roman soldiers in their service to the state.

"How glorious fall the valiant, sword in hand,
In front of battle for their native land!"[148]

—By Spartan general and poet Tyrtaeus, approximately the middle of the seventh century BCE

Swords and Warfare in the Classical World

Tetrarchs sculpture portraying detail of swords with hilts in the shape of eagles' heads, dating to the fourth century CE. The tetrarchs, from the Greek words for "Four rules," were the four co-rulers that governed the Roman Empire as long as Diocletian's reform lasted. Image source: Giovanni Dall'Orto, reproduced under Wikimedia Commons license.

NOTES

[1] See Nicholas Sekunda and Adam Hook, *Greek Hoplite, 480-323 BC: Weapons, Armor, Tactics* (Oxford, UK: Osprey Publishing, 2000), 3 & 9.

[2] Ibid., 16-17.

[3] Ibid., 9-10.

[4] Plutarch, *Moralia*, Vol. III of the Loeb Classical Library edition (1931). Plutarch was a Greek historian and essayist born in the first century CE.

[5] See Victor Davis Hanson, *Carnage and Culture: Landmark Battles in the Rise of Western Power* (New York, NY: Random House, 2001), 113-114 & 122.

[6] See Patrick Kelly, *Iron of the Empire: The History and Development of the Roman Gladius*, MyArmory.com, http://www.myarmoury.com/feature_ironempire.html. According to Sir Richard Francis Burton, a nineteenth century English explorer and soldier known for his research and writings about the sword, the word *xiphos* is used to describe a straight blade while the word *spati* is used to describe a saber-type sword. See Richard F. Burton, *The Book of the Sword* (Mineola, NY: Dover Publications, 1987), 235.

[7] See Rafael Trevino Martinez and Angus McBride, *Rome's Enemies: Spanish Armies, 218-19 BC* (Oxford, UK: Osprey Publishing, 1986), 20.

[8] See Ancient Edge, *Roman Gladius*, http://www.ancientedge.com/product_41_detailed.html.

[9] See Terence Wise and Richard Hook, *Armies of the Carthaginian Wars, 265-146 BC* (Oxford, UK: Osprey Publishing, 1982), 20.

[10] Polybius, *The Histories of Polybius*, Book III, translated by W. R. Paton, published in Vol. II of the Loeb Classical Library edition, 1922 through 1927.

[11] Ibid., Book II, published in Vol. I.

[12] See Titus Livius, *The History of Rome*, translated by D. Spillman (London, UK: Henry G. Bohn), 623. Note how safety was considered in extreme closeness rather than in a distance of a few feet.

[13] See Fernando Quesada Sanz, *Not So Different: Individual Fighting Techniques and Small Unit Tactics of Roman and Iberian Armies Within the Framework of Warfare in the Hellenistic Age*, Universidad Autónoma de Madrid, http://www.ffil.uam.es/equus/warmas/online/Not%20so%20different%20Quesada%20DEFINIT.pdf.

[14] See Hanson, *Carnage and Culture*, 119. The Punic Wars were fought between the Roman and Carthaginian empires for the primary purpose of securing control over Sicily, a territory that was important to the dominance of the Mediterranean Sea.

[15] See Polybius, Book III, published in Vol. II.

[16] Hanson, *Carnage and Culture*, 123 & 125.

[17] Ibid., 120.

[18] Ibid., 115-118.

[19] See Michael Sage, *Warfare in Ancient Greece: A Sourcebook* (New York, NY: Routledge, 1996), 19.

[20] See Bernhard Kauntz, *För 100 Generationer Sedan: År 550 f. Kr (100 Generations Ago: Year 550*

BC), http://www.werbeka.com/bibliote/500tal/550bc.htm.

[21] Richard Hooker, *The Celts* (1996), http://www.wsu.edu/~dee/MA/CELTS.HTM.

[22] John Boardman, et al., *The Oxford History of Greece and the Hellenistic World* (New York, NY: Oxford University Press, 1986), 261.

[23] See Anton Powell, *The Greek World* (New York, NY: Routledge, 1995), 28.

[24] Bruce Edward Blackistone, *Swords of Iron, Swords of Steel*, presented at the Miniver Cheevy Society for Early Medieval Studies (Mar. 1991), http://www.anvilfire.com/21centbs/armor/atli/swords1.htm.

[25] See Burton, 237.

[26] See Cornelius Nepos, *Lives of Eminent Commanders: Iphicrates*, translated by the Rev. John Selby Watson (1886), http://www.tertullian.org/fathers/nepos.htm#Iphicrates.

[27] See Burton, 238.

[28] See Nick Sekunda, *Greek Swords and Swordmanship*, Osprey Publishing, http://www.ospreypublishing.com/content2.php/cid=217.

[29] For comparison purposes, note that the Chinese philosopher Zhuangzi suggested around 300 BCE that it mattered little whether one chose to fight with a short or long sword. Rather, the best way to use a sword was to "make an empty feint. Then open your opponent by giving him an obvious advantage. Then strike, and get there first." See Zhuangzi, translated by Thomas Chen, China History Forum,

http://www.chinahistoryforum.com/lofiversion/index.php/t8919.html.

[30] See Nick Sekunda, *Greek Swords and Swordmanship* (Jan. 1, 2001), Osprey Publishing, http://www.ospreypublishing.com/articles/ancient_world/greek_swords_and_swordmanship.

[31] See Caius Julius Caesar, *De Bello Gallico & Other Commentaries of Caius Julius Caesar*, translated by W. A. Macdevitt (Everyman's Library, Project Gutenberg, 1929), 78-79.

[32] See Joseph Allen McCullough, *The Celtic Warrior in Britain: His Weapons, Equipment, and Armor* (Jun. 24, 2007), http://britishhistory.suite101.com/article.cfm/the_celtic_warrior_in_britain. "Archaeological evidence has proved that Celtic swords were of high quality, flexible and with a sharp, strong cutting edge, contradicting Polybius comments that in battle the blade quickly became so bent that the warrior had to straighten it with his foot. Confusion probably arose over the practice of ritually 'killing' a sword by deliberately bending it as part of a burial ceremony or sacrifice to the gods." See Stephen Allen, *Celtic Warrior, 300 BC-100 AD: Weapons, Armor, Tactics* (Oxford, UK: Osprey Publishing, 2001), 24.

[33] See Bertram Coghill Alan Windle, *Remains of the Prehistoric Age in England* (London, UK: Methuen & Co., 1904), 294-295.

[34] Ibid., 100.

[35] Miranda J. Green, *The Celtic World* (New York, NY: Routledge, 1996), 218. That the Celts had a warrior tradition is evident from burial finds, about half of which indicate that the warrior was buried with his weapons. Although the spear may have been

the primary Celtic weapon in the early tradition, nobles also carried a long sword with a blade about 36 inches in length. The earliest swords had definite points that enabled the warrior to slash or thrust with the weapon. Later swords indicate a rounded point that made the weapon nearly useless for thrusting. See Raimund Karl, *Ancient Celtic Warfare* (1999), Celtic Well, http://www.applewarrior.com/celticwell/ejournal/beltane/warfare.htm.

[36] See Peter Connolly and Hazel Dodge, *The Ancient City: Life in Classical Athens and Rome* (New York, NY: Oxford University Press, 1998), 214.

[37] See Robert Gunn, *Celtic History: The Ancient Celtic Warriors of Europe*, Scottish, Celtic, and Medieval History Online, http://members.aol.com/skyelander/menu14.html.

[38] See Allen, 24-25 & 61.

[39] Ibid., 22 & 25.

[40] See McCullough.

[41] See Karl.

[42] See Gunn.

[43] See Burton, 266-267.

[44] Ibid., 254-256.

[45] Ibid., 257-258.

[46] Ibid., 261. Burton also states that the people of northern Europe were superior in stature, and therefore favored the straight two-edged sword. See Burton, 127-128. The Roman swords later influenced the shape of the Viking swords, which were straight double edged weapons, often referred to as *vandil*, meaning "that which can be turned over" in the hand. Since the weapon was sharp on both edges, it could

be used with either hand and with either cutting edge facing the adversary. Vikings sometimes traded swords abroad or imported sword blades from France and fitted them with hilts made locally. The tenth century Arab chronicler Ibn Fadlan came in contact with the Vikings during their travels in Arab lands, and in his account of the Rus described their swords as broad, grooved, and of Frankish sort with blades tapering to a blunt tip, and used primarily for slashing or chopping rather than thrusting. See Martina Sprague, *Norse Warfare: Unconventional Battle Strategies of the Ancient Vikings* (New York, NY: Hippocrene Books, 2007), 145-147. The medieval longsword, too, was used in both thrusts and cuts, and would have required significant strength to wield successfully, particularly while wearing armor.

[47]See Flavius Vegetius Renatus, *The Military Institutions of the Romans*, translated from the Latin by John Clarke (1767), http://www.brainfly.net/html/books/brn0320.htm. Roman swords could both thrust and cut with efficiency, while Gallic swords were often considered of such poor quality that they could only cut. Even so, after only one downward cut, the Gallic swords tended to bend and had to be straightened "with a foot against the ground." See Janet Lang, "Study of the Metallography of Some Roman Swords," *Britannia*, Vol. 19 (1988), 199. Note that historians disagree about the "poor" quality of the Gallic swords.

[48]See Lang, 199.

[49]See Nic Fields, *Roman Auxiliary Cavalryman, AD 14-193* (Oxford, UK: Osprey Publishing, 2006), 16.

[50]See James R. Ross, et al., *Fighting Techniques of the Ancient World (3000 B.C. to 500 A.D.): Equipment, Combat Skills, and Tactics* (New York, NY: Thomas Dunne Books, 2002), 107-108.

[51]See Karen R. Dixon and Pat Southern, *The Roman Cavalry* (New York, NY: Routledge, 1992), 39 & 49.

[52]See Stephen Montague, *Ancient Armies: Spanish, 240-20 BC*, http://fanaticus.org/DBA/armies/dba52.html.

[53]See Jim Hrisoulas, "The Falcata–Historical Sword Spotlight," *Sword Forum Magazine Online* (Jan. 1999), http://swordforum.com/swords/historical/falcata.html.

[54]See Sanz. At first, it appears as though Iberian warriors differed little from their Roman counterparts. Neither weapons and armor, nor tactics differentiated the Iberian warriors significantly from the Romans. Rather, it was the organizational skill of the Romans that created determination and therefore an ability to fight using shock tactics, even when the odds did not seem in their favor. When the Roman armies were fighting far away from home, they still realized that they were fighting for Rome which gave them the morale and determination needed to emerge victorious.

[55]Ibid.

[56]See Robert L. O'Connell, *Of Arms and Men: A History of War, Weapons, and Aggression* (New York, NY: Oxford University Press, 1989), 68-69.

[57]See Richard Cohen, *By the Sword: A History of Gladiators, Musketeers, Samurai, Swashbucklers, and Olympic Champions* (New York, NY: Modern Library, 2002), 4-6.

[58] See Burton, 252-254. The gladiators fought each other in hand-to-hand combat and should not be confused with those fighters who combated a variety of wild beasts, a spectacle which developed in modern day into the bullfight.

[59] See Cohen, 9.

[60] See Vegetius. The author is doubtful whether it really benefits the soldier to exercise with a weapon that is heavier than the one used in the actual fight. The author, through her own experimentation, has found that such exercise tends to be detrimental to developing an acute sense of timing for attacks, defenses, and countercuts.

[61] Ibid.

[62] Polybius, Book X, published in Vol. IV.

[63] See Vegetius.

[64] See Dixon and Southern, 125 & 133.

[65] See John Keegan, *A History of Warfare* (New York, NY: Vintage Books, 1993), 242.

[66] Thucydides, *The Peloponnesian War: The Complete Hobbes Translation*, with notes and introduction by David Grene (Chicago, IL: University of Chicago Press, 1989), 13. Thucydides spends much time deliberating about war rather than discussing the actual fighting. His account stands in stark contrast to Julius Caesar's description of the war against Gaul, which details fighting strategy and tactics, the building of fortifications, etc.

[67] Ibid., 3-4 & 37.

[68] Ibid., 46.

[69] Ibid., 7 & 67-68.

[70] Plutarch, *Moralia*, Vol. III.

[71] See Sekunda, *Greek Swords and Swordmanship*.

[72] See Plutarch, *The Life of Alexander*, Vol. III of the Loeb Classical Library edition (1919).

[73] See Victor Davis Hanson, *The Western Way of War: Infantry Battle in Classical Greece* (Berkeley, CA: University of California Press, 2000), 165.

[74] See Sekunda, *Greek Swords and Swordmanship*.

[75] See History World, *Greek Citizen Armies: From the 7th Century BC*, http://www.historyworld.net/wrldhis/PlainTextHistories.asp?ParagraphID=bhu#292.

[76] See Hanson, *The Western Way of War*, 165. The phalanx allowed the warriors to launch a mass attack with considerable momentum behind it, while overcoming fear through peer pressure; they were fighting not as individuals but from a position of unified strength. There was little opportunity for flight. No soldier wanted to be perceived as a coward. The fear of loss of face was greater than the fear of death. The first rank of warriors would lead subsequent ranks and rely on their support. Although only the front rank could see the approaching enemy clearly, the rear ranks might get an indication of what was happening through a variety of sounds and the general movement of the formation. If a warrior fell, he would risk getting trampled by his peers or cut to death by his enemies. Movement must therefore be coordinated so that the warriors could give each other physical as well as emotional support. To prove effective, a phalanx several lines deep must rely on group cohesion and coordinated activity.

[77] See Thucydides, *History of the Peloponnesian War*, translated by Richard Crawley, http://history.eserver.org/peloponesian-war.txt.

[78] Harry Sidebottom, *Ancient Warfare: A Very Short Introduction* (New York, NY: Oxford University Press, 2004), 5-6. The Persians have been criticized by some military historians, most notably Victor Davis Hanson, to have, "suffered from that most dangerous tendency in war: a wish to kill but not to die in the process." See Keegan, 253. Hanson maintains that the West has enjoyed superiority in warfare due to certain cultural qualities, such as organization, discipline, morale, initiative, flexibility, and free inquiry and dissemination of knowledge. See Hanson, *Carnage and Culture*, 21. Note that other historians place greater emphasis on geographical location, terrain, and type of enemy threat when attempting to determine why the development of military tactics, strategy, and weapons varied in different parts of the world. See *Guns, Germs, and Steel: The Fates of Human Societies*, by Jared Diamond for a most intriguing study of why history proceeded in distinct ways for peoples of different parts of the globe.

[79] Sidebottom, 12-13.

[80] Ibid., 9.

[81] Ibid., 124.

[82] See Herodotus, *The Persian Wars*, translated by George Rawlinson (1942), http://www.parstimes.com/history/herodotus/persian_wars/thalia.html.

[83] See Diodorus Siculus, *The Historical Library of Diodorus Siculus*, translated by G. Booth (London, UK: W. McDowall, Pemberton Row, 1814),

386. Diodorus Siculus' writings cover Egyptian, Mesopotamian, Indian, Scythian, Arabian, and North African history, as well as parts of Greek and Roman history.

[84]See Robert E. Gaebel, *Cavalry Operations in the Ancient Greek World* (Norman, OK: University of Oklahoma Press, 2002), 29-31, 168 & 208.

[85]Xenohon, *Anabasis: The March Up Country*, translated by H. G. Dakyns (1998), Internet Ancient History Sourcebook, http://www.fordham.edu/halsall/ancient/xenophon-anabasis.html#BOOK%20II. Although Greece consists of numerous peninsulas and islands which complicated the use of foot combat, cavalry was not the primary force of the Greek armies. See Sidebottom, 89.

[86]Sidebottom, 14. Cavalry warfare and a desire to avoid hand-to-hand battle are, according to some historians, typical of the Middle Eastern tradition. See Keegan, 263.

[87]John A. Lynn, *Battle: A History of Combat and Culture* (Boulder, CO: Perseus Books, 2003), 10. For the purpose of this discussion, when talking about decisive battle (or decisive victory), it is assumed to mean that both belligerents understood, at the end of the battle, who was the victor. Historians differ in their definition of decisive battle. Some hold it to mean that the battle "decided something," regardless of whether or not that "something" was actually the preferred outcome. For example, the Korean War of 1950-1953 was decisive because it "decided that forcible unification of the peninsula was not attainable at [a] bearable cost to either side." See Colin S. Gray, *Defining and Achieving Decisive*

Victory, Strategic Studies Institute, U.S. Army War College (Apr. 2002), 8.
[88] See History World.
[89] See Burton, 268.
[90] Vegetius.
[91] See Sanz.
[92] See Roman Military Equipment, *The Dagger/Pugio*, http://www.romancoins.info/MilitaryEquipment-pugio.html. The daggers displayed on this Web site are from a variety of museums including the Leiden Museum in the Netherlands, the Landesmuseum in Bonn, the Archaeological Museum in Munich, and the Museum Carnuntinum in Austria.
[93] See Charlton T. Lewis and Charles Short, *A Latin Dictionary*, The Perseus Digital Library, http://www.perseus.tufts.edu/cgi-bin/ptext?layout.refdoc=Perseus%3Atext%3A1999.0 4.0059&layout.refcit=&doc=Perseus%3Atext%3A19 99.04.0059%3Aentry%3D%2339464&layout.reflook up=Pugio&layout.reflang=l.
[94] See The Ancient Library, *Dictionary of Greek and Roman Antiquities*, edited by William Smith (1870), 975.
[95] See Burton, 245.
[96] See Polybius, Book VI, published in Vol. III.
[97] Vegetius.
[98] See Burton, 259-261.
[99] See Polybius, Book XV, published in Vol. IV.
[100] Ibid., Book XVIII, published in Vol. V. A tightly packed formation was beneficial because of its momentum when it charged and the protection it

afforded the individual soldier. But it could also be detrimental to the warriors. If a formation was packed too tightly, it was difficult for the soldiers to maneuver individually. Each soldier could not swing his weapon with the same ease he could in a looser battle order. For greatest efficiency, a formation had to be tight when it charged, yet allow for more space between soldiers after the clash had occurred.

[101] See Dixon and Southern, 91.

[102] See Polybius, Book VI, published in Vol. III.

[103] See Caesar, 124.

[104] Ibid., 98.

[105] Ibid., 115. Caesar's reputation as a strong military leader and strategist is clearly displayed in his commentaries; although, he appears rather arrogant even toward his own troops.

[106] Ibid., 21.

[107] Ibid., 79. Titus Pulfio (Pullo) and Lucius Varenus were the main characters in the 2005-2006 historical drama television series *Rome*.

[108] Ibid., 32.

[109] Ibid., 136 & 160-161. Caesar went to war against his old ally Pompey the Great and the political rivals of the Roman senate who presented a threat to his dictatorship.

[110] Ibid., 41-42. The text is written by Caesar himself, probably in an attempt to glorify his own endeavors. Although no doubt biased, the account of his conquest of Gaul is an interesting source of historical information.

[111] Ibid., 167.

[112] See Livius, 663-664.

[113] See Dixon and Southern, 137.

[114]Ibid., 63.

[115]See K. M. Gilliver, *Caesar's Gallic Wars, 58-50 BC* (Oxford, UK: Osprey Publishing, 2003), 29.

[116]See Joe Zentner, "Adrianople: Last Great Battle of Antiquity," *Military History* (Oct. 2005), 56-60. When armor became scarce because of lack of funds, swords had to be made longer to compensate for the lack of sufficient armor. Note that the longsword, which became popular during the time of the medieval knight, was wielded with two hands, and the warrior generally did not carry a shield. The sword now acted as an instrument of both offense and defense.

[117]Ibid., 60.
[118]See Allen, 46 & 61.
[119]Green, 399.
[120]See Allen, 61.

[121]See J. Christopher Amberger, "Renaissance Sword Techniques in A.D. 150!" excerpted from *The Secret History of the Sword*, http://www.swordhistory.com/excerpts/greek.html.

[122]See Hanson, *The Western Way of War*, 75.

[123]See James Scott Wheeler, "Armor," *The Reader's Companion to Military History*, edited by Robert Cowley and Geoffrey Parker (New York, NY: Houghton Mifflin Company, 1996), 31.

[124]See Keegan, 250.
[125]See Hanson, *The Western Way of War*, 56.

[126]A thousand years later, the Norse warriors, too, despite the often cooler temperatures of their raiding grounds, had it as a habit to rid themselves of heavy armor when hot, tired, wounded, or returning

to the ship after a successful raid. See Sprague, 159-160.

[127] Hanson, *The Western Way of War*, 35 & 83.

[128] Xenophon.

[129] See Plutarch, *The Life of Alexander*, Vol. III. Plutarch wrote almost four hundred years after Alexander's death, but justifies his writings on the account that "it is not Histories that I am writing, but Lives; and in the most illustrious deeds there is not always a manifestation of virtue or vice, nay, a slight thing like a phrase or a jest often makes a greater revelation of character than battles when thousands fall, or the greatest armaments, or sieges of cities. Accordingly, just as painters get the likenesses in their portraits from the face and the expression of the eyes, wherein the character shows itself, but make very little account of the other parts of the body, so I must be permitted to devote myself rather to the signs of the soul in men, and by means of these to portray the life of each, leaving to others the description of their great contests."

[130] Ibid.

[131] Ibid.

[132] See Herodotus.

[133] See Wheeler, 31.

[134] See Fields, 14.

[135] See Gunn.

[136] See Victor Davis Hanson, "Legions," *The Reader's Companion to Military History*, edited by Robert Cowley and Geoffrey Parker (New York, NY: Houghton Mifflin Company, 1996), 260.

[137] See Cohen, 9-10.

[138] Plutarch, *Caesar*, translated by John Dryden, The Internet Classics Archive, http://classics.mit.edu/Plutarch/caesar.html.

[139] See Kelly.

[140] Ibid. Note that the gladius came in several subtle variations. The type known as Mainz (the name refers to how and where the weapon was found) was slightly longer, up to 22 inches in length with a long sharp point, and proved an effective weapon for thrusting through the armor worn by Rome's enemies. The type known as Pompeii was a bit shorter around 19 inches in length with a short and nearly triangular point. The simpler Pompeii sword with straight edges was easier and less expensive to manufacture, and large numbers of troops could thus be equipped.

[141] See Kelly.

[142] See Dixon and Southern, 105. By contrast, according to Plutarch's writings about the life of Caesar and Pompey, when Pompey threatened to use a sword and buckler against his enemies, the nobles much resented this, thinking it unsuitable for a man of his status and was unbecoming "the reverence due to the senate, but resembling rather the vehemence of a boy or the fury of a madman." See Plutarch, *Caesar*. The value one placed on the sword as a battlefield weapon and to honor the gods is also evident in Herodotus' account of the Scythian sacrifices of wild animals, and the "rites paid to Mars." On a great pile of 150 wagon loads of brushwood is planted an antique iron sword, which serves as the image of Mars: "[Y]early sacrifices of cattle and of horses are made to it, and more victims are offered thus than to all the rest of their gods." See Herodotus.

[143] See Burton, 252.

[144] See Josiah Ober, "Spartans," *The Reader's Companion to Military History*, edited by Robert Cowley and Geoffrey Parker (New York, NY: Houghton Mifflin Company, 1996), 438.

[145] See Tyrtaeus, *Hoplite Revolution*, translated by Walter Donlan, http://www.msu.edu/~tyrrell/hoplite.htm.

[146] See Sage, xv.

[147] See Siculus, 444-445.

[148] Tyrtaeus, *Martial Elegy*, Poetry Archive, http://www.poetry-archive.com/t/tyrt%E6us.html.

BIBLIOGRAPHY

Allen, Stephen. *Celtic Warrior, 300 BC-100 AD: Weapons, Armor, Tactics.* Oxford, UK: Osprey Publishing, 2001.

Amberger, J. Christopher. "Renaissance Sword Techniques in A.D. 150!" Excerpted from *The Secret History of the Sword.* http://www.swordhistory.com/excerpts/greek.html.

Ancient Edge. *Roman Gladius.* http://www.ancientedge.com/product_41_detailed.html.

Blackistone, Bruce Edward. *Swords of Iron, Swords of Steel.* Presented at the Miniver Cheevy Society for Early Medieval Studies (Mar. 1991). http://www.anvilfire.com/21centbs/armor/atli/swords1.htm.

Boardman, John, et al. *The Oxford History of Greece and the Hellenistic World.* New York, NY: Oxford University Press, 1986.

Burton, Richard F. *The Book of the Sword.* Mineola, NY: Dover Publications, 1987.

Caesar, Caius Julius. *De Bello Gallico & Other Commentaries of Caius Julius Caesar.* Translated by W. A. Macdevitt. Everyman's Library, Project Gutenberg, 1929.

Cohen, Richard. *By the Sword: A History of Gladiators, Musketeers, Samurai, Swashbucklers, and Olympic Champions.* New York, NY: Modern Library, 2002.

Connolly, Peter and Dodge, Hazel. *The Ancient City: Life in Classical Athens and Rome.* New York, NY: Oxford University Press, 1998.

Dixon, Karen R. and Southern, Pat. *The Roman Cavalry.* New York, NY: Routledge, 1992.

Fields, Nic. *Roman Auxiliary Cavalryman, AD 14-193.* Oxford, UK: Osprey Publishing, 2006.

Gaebel, Robert E. *Cavalry Operations in the Ancient Greek World.* Norman, OK: University of Oklahoma Press, 2002.

Gilliver, K. M. *Caesar's Gallic Wars, 58-50 BC.* Oxford, UK: Osprey Publishing, 2003.

Gray, Colin S. *Defining and Achieving Decisive Victory.* Strategic Studies Institute, U.S. Army War College (Apr. 2002).

Green, Miranda J. *The Celtic World.* New York, NY: Routledge, 1996.

Gunn, Robert. *Celtic History: The Ancient Celtic Warriors of Europe.* Scottish, Celtic, and Medieval History Online; http://members.aol.com/skyelander/menu14.html.

Hanson, Victor Davis. *Carnage and Culture: Landmark Battles in the Rise of Western Power*. New York, NY: Random House, 2001.

………"Legions." *The Reader's Companion to Military History.* Edited by Robert Cowley and Geoffrey Parker. New York, NY: Houghton Mifflin Company, 1996.

………*The Western Way of War: Infantry Battle in Classical Greece.* Berkeley, CA: University of California Press, 2000.

Herodotus. *The Persian Wars.* Translated by George Rawlinson (1942). http://www.parstimes.com/history/herodotus/persian_wars/thalia.html.

History World. *Greek Citizen Armies: From the 7th Century BC.* http://www.historyworld.net/wrldhis/PlainTextHistories.asp?ParagraphID=bhu#292.

Hooker, Richard. *The Celts* (1996). http://www.wsu.edu/~dee/MA/CELTS.HTM.

Hrisoulas, Jim. "The Falcata–Historical Sword Spotlight." *Sword Forum Magazine Online* (Jan. 1999). http://swordforum.com/swords/historical/falcata.html.

Karl, Raimund. *Ancient Celtic Warfare* (1999). Celtic Well.

http://www.applewarrior.com/celticwell/ejournal/beltane/warfare.htm.

Kauntz, Bernhard. *För 100 Generationer Sedan: År 550 f. Kr (100 Generations Ago: Year 550 BC).* http://www.werbeka.com/bibliote/500tal/550bc.htm.

Keegan, John. *A History of Warfare.* New York, NY: Vintage Books, 1993.

Kelly, Patrick. *Iron of the Empire: The History and Development of the Roman Gladius.* MyArmory.com. http://www.myarmoury.com/feature_ironempire.html.

Lang, Janet. "Study of the Metallography of Some Roman Swords." *Britannia*, Vol. 19 (1988).

Lewis, Charlton T. and Short, Charles. *A Latin Dictionary.* The Perseus Digital Library. http://www.perseus.tufts.edu/cgi-bin/ptext?layout.refdoc=Perseus%3Atext%3A1999.04.0059&layout.refcit=&doc=Perseus%3Atext%3A1999.04.0059%3Aentry%3D%2339464&layout.reflookup=Pugio&layout.reflang=l.

Livius, Titus. *The History of Rome.* Translated by D. Spillman. London, UK: Henry G. Bohn.

Lynn, John A. *Battle: A History of Combat and Culture.* Boulder, CO: Perseus Books, 2003.

Martinez, Rafael Trevino and McBride, Angus. *Rome's Enemies: Spanish Armies, 218-19 BC.* Oxford, UK: Osprey Publishing, 1986.

McCullough, Joseph Allen. *The Celtic Warrior in Britain: His Weapons, Equipment, and Armor* (Jun. 24, 2007). http://britishhistory.suite101.com/article.cfm/the_celtic_warrior_in_britain.

Montague, Stephen. *Ancient Armies: Spanish, 240-20 BC.* http://fanaticus.org/DBA/armies/dba52.html.

Nepos, Cornelius. *Lives of Eminent Commanders: Iphicrates.* Translated by the Rev. John Selby Watson (1886). http://www.tertullian.org/fathers/nepos.htm#Iphicrates.

Ober, Josiah. "Spartans." *The Reader's Companion to Military History.* Edited by Robert Cowley and Geoffrey Parker. New York, NY: Houghton Mifflin Company, 1996.

O'Connell, Robert L. *Of Arms and Men: A History of War, Weapons, and Aggression.* New York, NY: Oxford University Press, 1989.

Plutarch. *Caesar.* Translated by John Dryden. The Internet Classics Archive. http://classics.mit.edu/Plutarch/caesar.html.

………*Moralia.* Vol. III of the Loeb Classical Library edition (1931).

………*The Life of Alexander.* Vol. III of the Loeb Classical Library edition (1919).

Polybius. *The Histories of Polybius.* Book III. Translated by W. R. Paton. Published in Vol. II of the Loeb Classical Library edition, 1922 through 1927.

Powell, Anton. *The Greek World.* New York, NY: Routledge, 1995.

Renatus, Flavius Vegetius. *The Military Institutions of the Romans.* Translated from the Latin by John Clarke (1767). http://www.brainfly.net/html/books/brn0320.htm.

Roman Military Equipment. *The Dagger/Pugio.* http://www.romancoins.info/MilitaryEquipment-pugio.html.

Ross, James R. et al. *Fighting Techniques of the Ancient World (3000 B.C. to 500 A.D.): Equipment, Combat Skills, and Tactics.* New York, NY: Thomas Dunne Books, 2002.

Sage, Michael. *Warfare in Ancient Greece: A Sourcebook.* New York, NY: Routledge, 1996.

Sanz, Fernando Quesada. *Not So Different: Individual Fighting Techniques and Small Unit Tactics of Roman and Iberian Armies Within the Framework of Warfare in the Hellenistic Age.* Universidad Autónoma de Madrid. http://www.ffil.uam.es/equus/warmas/online/Not%20so%20different%20Quesada%20DEFINIT.pdf.

Sekunda, Nicholas and Hook, Adam. *Greek Hoplite, 480-323 BC: Weapons, Armor, Tactics*. Oxford, UK: Osprey Publishing, 2000.

Sekunda, Nick. *Greek Swords and Swordmanship.* Osprey Publishing. http://www.ospreypublishing.com/content2.php/cid=217.

.... *Greek Swords and Swordmanship* (Jan. 1, 2001). Osprey Publishing. http://www.ospreypublishing.com/articles/ancient_world/greek_swords_and_swordmanship.

Siculus, Diodorus. *The Historical Library of Diodorus Siculus.* Translated by G. Booth. London, UK: W. McDowall, Pemberton Row, 1814.

Sidebottom, Harry. *Ancient Warfare: A Very Short Introduction.* New York, NY: Oxford University Press, 2004.

Sprague, Martina. *Norse Warfare: Unconventional Battle Strategies of the Ancient Vikings.* New York, NY: Hippocrene Books, 2007.

The Ancient Library. *Dictionary of Greek and Roman Antiquities.* Edited by William Smith (1870).

Thucydides. *History of the Peloponnesian War.* Translated by Richard Crawley. http://history.eserver.org/peloponesian-war.txt.

.........*The Peloponnesian War: The Complete Hobbes Translation.* With notes and introduction by David Grene. Chicago, IL: University of Chicago Press, 1989.

Tyrtaeus. *Hoplite Revolution.* Translated by Walter Donlan. http://www.msu.edu/~tyrrell/hoplite.htm.

.........*Martial Elegy.* Poetry Archive. http://www.poetry-archive.com/t/tyrt%E6us.html.

Wheeler, James Scott. "Armor." *The Reader's Companion to Military History.* Edited by Robert Cowley and Geoffrey Parker. New York, NY: Houghton Mifflin Company, 1996.

Windle, Bertram Coghill Alan. *Remains of the Prehistoric Age in England.* London, UK: Methuen & Co., 1904.

Wise, Terence and Hook, Richard. *Armies of the Carthaginian Wars, 265-146 BC.* Oxford, UK: Osprey Publishing, 1982.

Xenohon. *Anabasis: The March Up Country.* Translated by H. G. Dakyns (1998). Internet Ancient History Sourcebook. http://www.fordham.edu/halsall/ancient/xenophon-anabasis.html#BOOK%20II.

Zentner, Joe. "Adrianople: Last Great Battle of Antiquity." *Military History* (Oct. 2005).

Zhuangzi. Translated by Thomas Chen. China History Forum.
http://www.chinahistoryforum.com/lofiversion/index.php/t8919.html.

About the Author:

Martina Sprague has a Master of Arts Degree in Military History from Norwich University in Vermont. She is the author of numerous books about military and general history. For more information, please visit her Web site: www.modernfighter.com.

Made in the USA
Middletown, DE
29 September 2018